智能微型运动装置（Micromouse）技术与应用系列丛书
天津市"一带一路"联合实验室（研究中心）项目研究成果
工程实践创新项目（EPIP）教学模式规划教材

# 智能鼠原理与制作
## （高级篇）

ZHINENGSHU YUANLI YU ZHIZUO（GAOJI PIAN）

王 超 高 艺 宋立红 编著
周蘩雨 严靖怡 刘 佳 编译

U0183852

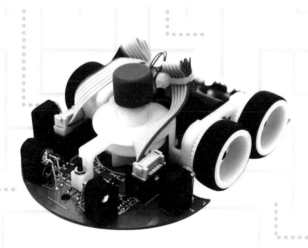

中国铁道出版社有限公司
CHINA RAILWAY PUBLISHING HOUSE CO., LTD.

# 内 容 简 介

本书为中英双语版，以天津启诚伟业科技有限公司提供的 TQD-Micromouse-JM 智能鼠为载体，是智能微型运动装置（Micromouse）技术与应用系列丛书的高级篇。

本书以真实工程项目为背景，通过"基础知识篇"、"综合实践篇"和"拓展竞技篇"三篇讲述了智能鼠的发展、硬件、开发环境、基本操作；智能鼠的运动姿态控制、智能控制算法和技术；智能鼠迷宫信息的获取与存储、路径规划、走迷宫程序设计等。同时，本书附录提供了国际 Micromouse 走迷宫竞赛相关知识、智能鼠器件清单、智能鼠迷宫图库、专业词汇中英对照表、国际实训课程标准等丰富资源。

本书在重要的知识点、能力点和素养点上，配有丰富的视频、图片、文本等资源，学习者可以通过扫描书中二维码获取相关信息。

本书适合作为高等院校，特别是"新工科"专业信息与自动化技术融合与创新学科教学的指导用书，还可作为相关工程技术人员培训用书及智能鼠爱好者参考用书。

**图书在版编目（CIP）数据**

智能鼠原理与制作 . 高级篇：汉、英／王超，高艺，宋立红编著 . —北京：中国铁道出版社有限公司，2021.3
（智能微型运动装置（Micromouse）技术与应用系列丛书）
ISBN 978-7-113-27550-1

Ⅰ . ①智… Ⅱ . ①王… ②高… ③宋… Ⅲ . ①智能机器人－程序设计－汉、英 Ⅳ . ① TP242.6

中国版本图书馆 CIP 数据核字（2021）第 035147 号

书　　名：**智能鼠原理与制作（高级篇）**
作　　者：王　超　高　艺　宋立红

策　　划：何红艳　　　　　　　　　　编辑部电话：（010）83552550
责任编辑：何红艳　绳　超
封面设计：刘　颖
责任校对：焦桂荣
责任印制：樊启鹏

出版发行：中国铁道出版社有限公司（100054，北京市西城区右安门西街 8 号）
网　　址：http://www.tdpress.com/51eds/
印　　刷：北京柏力行彩印有限公司
版　　次：2021 年 3 月第 1 版　2021 年 3 月第 1 次印刷
开　　本：787 mm×1 092 mm　1/16　印张：18　字数：302 千
书　　号：ISBN 978-7-113-27550-1
定　　价：66.00 元

**王 超**

天津大学电气自动化与信息工程学院教授，教育部高等学校自动化类专业教学指导委员会委员，主要从事多相流检测与仪器和电学层析成像的研究（ERT，ECT，EMT 和 EST），在天津大学教授计算控制技术和工业控制网络课程，自 2010 年起，首次将智能鼠作为重要的实践教学载体引入电气自动化与信息工程学院。2018 年，在第 33 届 APEC 国际电脑鼠竞赛中，天津大学的两支队伍包揽冠亚军。

**高 艺**

南开大学电子信息工程学院硕士生导师，电子信息实验教学中心副主任，天津市单片机学会青年骨干工作委员会副主任，多项天津市大学生竞赛及职业技能竞赛裁判组成员。先后参与多项"国家高技术研究发展计划（863 计划）项目"、"天津市科技支撑计划重点项目"以及横向科研项目。多次作为指导教师带队参加全国大学生电子设计竞赛、天津市电子设计竞赛、天津市物联网竞赛、天津市大学生 IEEE 电脑鼠竞赛、APEC 国际电脑鼠竞赛、全国机器人大赛等。

**宋立红**

天津启诚伟业科技有限公司总经理，启诚智能鼠创始人。多年来专注于高等教育、职业教育、基础教育领域的嵌入式、物联网、人工智能等教学仪器设备研发、设计、生产、推广、服务工作。40 余次赞助支持大学生学科竞赛"启诚杯"智能鼠走迷宫赛项及职业院校技能竞赛智能微型运动装置（智能鼠）赛项等。从 2016 年开始，积极致力于国际项目鲁班工坊技术支持服务工作，启诚智能鼠作为中国创新型教育装备，伴随鲁班工坊不远万里前往泰国、印度、印尼、巴基斯坦、柬埔寨、尼日利亚、埃及等国家，受到所在国师生一致青睐，为"一带一路"倡议做出了努力和贡献。

### 周蘩雨

美国中央俄克拉荷马大学英语教育学硕士，现任职于天津机电职业技术学院国际交流与合作处。在美期间参与了中英语言对比研究等研究项目，在孔子课堂教授了大批以英语为第二语言的非美籍学生，从中获取了大量教学经验。2018年参加夏季达沃斯论坛，为国际能源论坛秘书长孙贤胜做专属随行翻译及事务官。2018年参与葡萄牙鲁班工坊创设工作；2018年发表《"鲁班工坊"教学模式在电脑鼠实践教学中的应用》中英双语论文。2019年参与马达加斯加鲁班工坊创设工作，其间完成了多场会议翻译、随行翻译及笔译，累计笔译文件及材料超过20万字。

### 严靖怡

天津启诚伟业科技有限公司总经理助理，就读于美国加州大学圣克鲁兹分校。2015年担任美国APEC智能鼠国际竞赛组委会主席MIT David Otten教授在中国交流访问期间随行翻译。2018年担任新蒙古教育集团董事会副主席Davaanyam访问天津智能鼠竞赛交流活动随行英语翻译及会议同声传译。2018年自费到柬埔寨做志愿者，担任柬埔寨国立理工学院Bun Phearin校长鲁班工坊智能鼠项目翻译，帮助柬埔寨学生学习中国智能鼠技术，为"一带一路"沿线国家教育发展做出了努力和贡献。

### 刘 佳

南开大学外国语学院公共英语教学部讲师，南开大学外国语学院博士生。国家级精品课"大学英语"和国家级精品课"科研方法论"主讲教师。曾两次被评为南开大学外国语学院公共英语教学部优秀教师。曾荣获南开大学教学基本功大赛一等奖，天津市教学基本功大赛二等奖。主持和参与多项市级和校级科研项目，发表数十篇科研论文。参编口语教材、翻译资格考试教材、写作教材和词典等数本，翻译原版英文书籍数本。

扫码观看

智能鼠名人榜

——Mr. David Michael Otten

I am very pleased to find out that you are going to write a book about micromouse. This contest is a fantastic way to learn about electromechanical systems and integrating hardware and software. I have learned a great deal during my 30 years with this contest and I am sure your readers will also. Congratulations and good luck with this endeavor.

David Otten
APEC micromouse chair

我很高兴得知你们要编写关于智能鼠的书。智能鼠大赛是学习机电系统和集成软硬件的绝佳方式。在过去经历的30年比赛中，我学到了很多东西，我相信你们的读者也会这样。祝贺你们并祝好运！

## Mr. David Michael Otten

美国麻省理工学院高级研发工程师。
国际智能鼠走迷宫教育教学专家。
多年从事AI机器人开发研究工作，连续30余届美国APEC国际电脑鼠竞赛组委会主席，曾多次参加日本、新加坡、美国、英国等国家智能鼠比赛，并多次蝉联世界冠军。

扫码观看

智能鼠名人榜

——Mr. Peter Harrison

The micromouse contest is an integration of multiple disciplines and many technologies. It involves machine engineering, electronic engineering, automatic control, artificial intelligence, program design, sensing and testing technology

The micromouse contest will enhance the participant's technology level and application abilities, providing a platform for technological innovation

The publication of books on micromouse education will play a significant role in learning micromouse technology for Chinese students. The micromouse robots made by Qicheng are to the world through Luban workshops, benifitting students around the world

Congratulations on the publication of the micromouse book series. They provide convenience and reference for micromouse fans and students at all levels.

智能鼠是集多学科多技术的融合体，主要涉及机械工程、电子工程、自动控制、人工智能、程序设计、传感与测试技术等学科。

竞赛的开展，必然提升参赛者在相关领域的技术水平和应用能力，为技术创新提供平台。

这类教育资源书籍的出版，对于中国乃至世界学生学习智能鼠技术都有着重要的意义和作用，启诚电脑鼠通过鲁班工坊冲出国门走向世界，让世界各地的学生受益，我为启诚智能鼠感到骄傲和自豪！

由衷地祝贺智能微型运动装置（Micromouse）技术与应用系列丛书出版。这为智能鼠爱好者、不同教育层面的学生学习嵌入式微型机器人（智能鼠）提供了便利和参考。

## Mr. Peter Harrison

英国伯明翰城市大学高级研发工程师。

国际智能鼠走迷宫教育教学专家。

多年来从事设计、研发IT集成项目工作，在培养大学生人工智能机器人技术领域、实训教学和实用技能方面成绩卓著，曾多次参加美国、日本、新加坡、英国等国家的智能鼠走迷宫比赛，并多次蝉联世界冠军。

When about ten years ago I started organizing robotics events with the Micromouse contest, I did it because I realized that this robotics contest would be the most applicable to encourage students to STEM areas. It is a complete contest that involves students of all ages and skill levels. However, there is still very little aggregated information on this topic. Thus, the publication of a series of books on Micromouse is beneficial.

It is with great pleasure that I, as an enthusiast and as the organizer of the Micromouse Portuguese Contest, give my support to that series of publications on Micromouse, being certain that they will contribute to the enhancement of Micromouse.

*Antonio Luis Gomes Valente*

---

十年前，我开始组织智能鼠比赛，因为我认为这个比赛最适合激发学习者对 STEM 领域的热情。各个年龄和各种水平的选手都可以参加这个比赛。但遗憾的是，有关这个领域鲜有详细的介绍。因此，有关智能鼠的这套书十分有益。作为一个智能鼠爱好者和葡萄牙智能鼠赛事的组织者，我向大家推荐这套书。我认为这套书将会促进智能鼠的发展。

---

## Mr. António Valente

葡萄牙 Trás-os-Montes and Alto Douro 大学科学技术学院教授，高级研究员。

葡萄牙国际智能鼠大赛组委会主席。

研究方向：MEMS 传感器、微控制器和嵌入式系统，重点是农业应用。完成了多项葡萄牙国家级和国际级资助的纵向和横向科研项目和课题，包括：安全、Eno-分析、RobTech、IPAVPSI、Focus 等。

　　"智能鼠"，英文名为 Micromouse，是使用嵌入式微控制器、传感器和机电运动部件构成的一种智能微型运动装置（嵌入式微型机器人），智能鼠可以在不同"迷宫"中自动记忆和选择路径，采用相应的算法，快速到达所设定的目的地。智能鼠走迷宫竞赛结合了机械电子、控制、光学、程序设计和人工智能等多方面的科技知识。

　　四十多年来，电气电子工程师学会（IEEE）每年举办一次国际性的智能鼠走迷宫竞赛，自举办以来各国踊跃参加，尤其是美国和欧洲国家的高校学生，为此有些大学还特别开设了"智能鼠原理与制作"的选修课程。中国从2007年开始在上海长三角地区举行小规模尝试性比赛。2009年天津启诚伟业科技有限公司将这项国际赛事引进天津，以工程实践创新项目（EPIP）教学模式，对智能鼠走迷宫竞赛进行本土化创新改革，对于后期智能鼠竞赛的开展和走进课堂、融入教学起到关键性的作用。经过多年的蜕变与优化，"智能鼠"已经成为集"职业性、综合性、先进性、趣味性"于一体的创新实践教育平台，在推动课程改革，提高教学质量，培养学习者的工程实践创新能力等方面发挥了重要的作用。

　　为了将智能鼠的成果进一步推广应用，我们编写了适用于高等院校学生学习的《智能鼠原理与制作》（高级篇），本书以天津启诚伟业科技有限公司提供的 TQD-Micromouse-JM 智能鼠为载体，由浅及深、由易到难、循序渐进地进行教学。

　　本书遵循递进原则，从"玩转"到"掌握"，再到"精通"，丰富学习者的工程实践知识和技术应用经验，拓展学习者的专业视野，内化形成良好的职业素养，提升学习者的实践创新能力。本书所选案例均来自真实的工程项目，编者均来自国内长期从事智能鼠研究与开发、国际智能鼠走迷宫竞赛获奖的院校和企业。

　　本书在重要的知识点、能力点和素养点上，配有丰富视频、图片、文本等资源，学习者可以通过扫描书中二维码获取相关信息。本书编著者长期的国际化教学活动积淀，使得本书成为推进国际化人才培养的实践教学载体，智能微型运动装置（Micromouse）技术与应用系列丛书是天津市"一带一路"联合实验室（研究中心）——天津中德柬埔寨智能运动装置与互联通信技术推广中心研究成果，同时也是工程实践创新项目（EPIP）教学模式规划教材。

在本书附录中提供"智能鼠原理与制作"国际课程教学大纲，以专业基础课程为前导，学生通过十个项目的学习训练，可具备智能机器人硬件设计与驱动、软件设计与编程、项目工程实施等工程素养；深入理解传感器与检测信号调试、电动机精密控制、机器人智能搜索与路径规划等专业知识。本课程内容与多个国家的"鲁班工坊"建设项目高度融合，服务"一带一路"倡议，推广中国教育标准，为"一带一路"沿线国家提供丰富实践教学资源，服务各地技术技能人才培养。

本书由天津大学教授王超，南开大学副教授高艺，启诚智能鼠创始人、天津启诚伟业科技有限公司总经理宋立红编著。英文部分由天津机电职业技术学院讲师周蘩雨，天津启诚伟业科技有限公司总经理助理严靖怡，南开大学外国语学院公共英语教学部讲师刘佳编译，天津中德应用技术大学副教授张链、讲师高源参与了本书部分翻译工作。国际专家美国麻省理工学院教授David Otten和英国伯明翰城市大学教授Peter Harrison、葡萄牙Trás-os-蒙特斯与奥拓杜罗大学教授António Valente参与本书英文内容的译审，并专门为本书写了贺信。本书在编写过程中得到了天津大学、南开大学、天津中德应用技术大学、英国伯明翰城市大学和葡萄牙Trás-os-蒙特斯与奥拓杜罗大学等相关院校教授专家的大力支持。天津启诚伟业科技有限公司陈立考、邱建国、宋姗为本书出版提供了企业实际工程案例、二维码视频、动画、PPT等课程资源。衷心感谢天津市教育委员会、中国铁道出版社有限公司、天津启诚伟业科技有限公司对本教学资源开发提供的指导与帮助。中国铁道出版社有限公司支持出版，并通过鲁班工坊在"一带一路"沿线国家使用。

限于编著者的经验、水平以及时间，书中难免存在不妥和疏漏之处，敬请专家、广大读者批评指正。

编著者

2020年8月

# 第一篇 基础知识篇

　　智能鼠走迷宫竞赛在国际上已经有40多年的历史，竞赛要求智能鼠从起点出发，在不受人为操纵影响的条件下在未知的迷宫中，自主搜索迷宫找到终点，并挑选出最短的一条路径进行冲刺。竞赛成绩根据搜索迷宫的时间和冲刺时间区分名次，竞赛迷宫遵照电气电子工程师学会（IEEE）的国际标准。在本篇中，将分别从国际IEEE标准迷宫场地、智能鼠的硬件系统和软件开发环境等方面系统学习智能鼠技术，并对智能鼠的基本原理和实际操作方法进行具体说明。

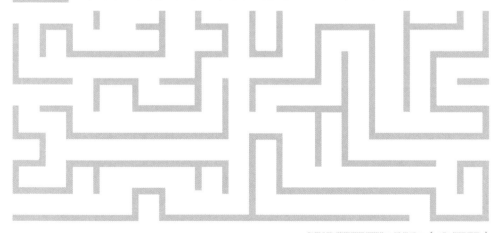

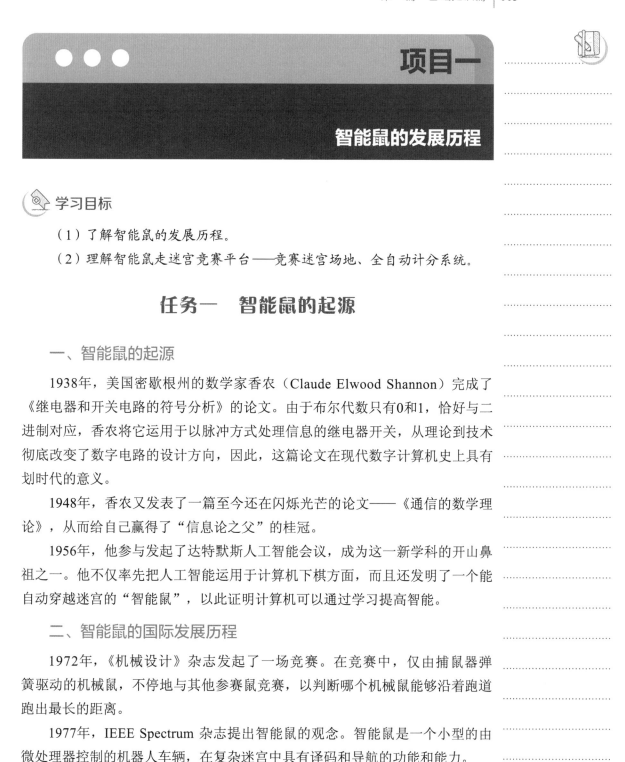

项目一

**智能鼠的发展历程**

### 学习目标

（1）了解智能鼠的发展历程。

（2）理解智能鼠走迷宫竞赛平台——竞赛迷宫场地、全自动计分系统。

# 任务一　智能鼠的起源

## 一、智能鼠的起源

1938年，美国密歇根州的数学家香农（Claude Elwood Shannon）完成了《继电器和开关电路的符号分析》的论文。由于布尔代数只有0和1，恰好与二进制对应，香农将它运用于以脉冲方式处理信息的继电器开关，从理论到技术彻底改变了数字电路的设计方向，因此，这篇论文在现代数字计算机史上具有划时代的意义。

1948年，香农又发表了一篇至今还在闪烁光芒的论文——《通信的数学理论》，从而给自己赢得了"信息论之父"的桂冠。

1956年，他参与发起了达特默斯人工智能会议，成为这一新学科的开山鼻祖之一。他不仅率先把人工智能运用于计算机下棋方面，而且还发明了一个能自动穿越迷宫的"智能鼠"，以此证明计算机可以通过学习提高智能。

## 二、智能鼠的国际发展历程

1972年，《机械设计》杂志发起了一场竞赛。在竞赛中，仅由捕鼠器弹簧驱动的机械鼠，不停地与其他参赛鼠竞赛，以判断哪个机械鼠能够沿着跑道跑出最长的距离。

1977年，IEEE Spectrum 杂志提出智能鼠的观念。智能鼠是一个小型的由微处理器控制的机器人车辆，在复杂迷宫中具有译码和导航的功能和能力。

1979年，电气电子工程师学会（IEEE）通过其Spectrum and Computer杂志

发起了一场智能鼠竞赛，奖励能够在最短时间内自主走出迷宫的智能鼠的设计者1 000美元。

1980年，东京举办了首场全日本Micromouse国际公开赛，之后，又有多个比赛被创办，如：1980年英国智能鼠大赛，1987年新加坡举办了第一届新加坡Micromouse竞赛和2007年中国计算机学会举办的首场IEEE国际标准Micromouse走迷宫竞赛等，如图1-1-1所示。

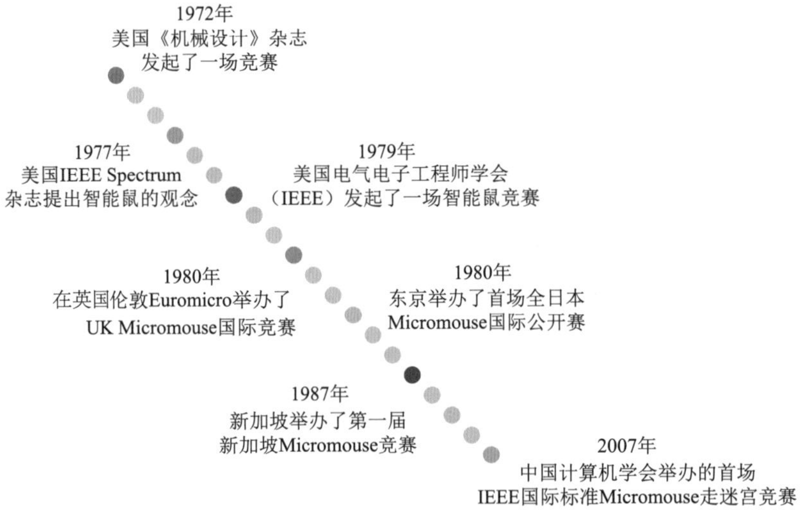

**1972年**
美国《机械设计》杂志
发起了一场竞赛

**1977年**
美国IEEE Spectrum
杂志提出智能鼠的观念

**1979年**
美国电气电子工程师学会
（IEEE）发起了一场智能鼠竞赛

**1980年**
在英国伦敦Euromicro举办了
UK Micromouse国际竞赛

**1980年**
东京举办了首场全日本
Micromouse国际公开赛

**1987年**
新加坡举办了第一届
新加坡Micromouse竞赛

**2007年**
中国计算机学会举办的首场
IEEE国际标准Micromouse走迷宫竞赛

图1-1-1　智能鼠国际发展

从最初1972年的机械电子鼠发展到现在的智能鼠，经过了40多年的蜕变，参加竞赛的选手从开始仅限于哈佛大学、麻省理工学院等世界知名学府的研究生，发展到从研究型大学到应用技术大学再到职业院校的学生，甚至是中小学生。多教育层面都采纳智能鼠作为教学载体，培养学生们的工程素养以及科技创新意识、动手设计能力。

各类智能鼠竞赛也如雨后春笋般蓬勃发展。目前智能鼠竞赛已经成为应用于不同教育阶段的国际创新型学生竞赛。

### 三、智能鼠的中国发展历程

从2007年至今，智能鼠在中国经历了十余年的发展历程，如图1-1-2所示。2007年天津启诚伟业科技有限公司将这项国际赛事引进天津，以中国先进的教育模式"工程实践创新项目"为核心理念，对智能鼠走迷宫竞赛进行本土化创新改革，助力智能鼠竞赛在中国的蓬勃开展，对于智能鼠技术走进课堂融入教学起到关键性的引领作用。

● 视频

中国智能鼠
发展

图1-1-2　智能鼠在中国的发展

　　竞赛对于满足产业优化升级，开阔国际视野，掌握实践与创新经验，培养高技术、高技能人才，起到了引领推动作用（见图1-1-3）。智能鼠在中国从大学生竞赛到职业院校大赛，再到普职融通国际挑战赛，积累了丰富的竞赛经验和优秀的技术积淀。

图1-1-3　竞赛纪实照片

　　十余年来，中国的智能鼠竞赛不断创新国际发展新思路，从最初的"简单模仿"学习，发展到目前的"互学互鉴"，逐步搭建起国际交流合作的新平台，先后经历了学习借鉴、蜕变升华和引领辐射三个阶段。

　　首先是学习借鉴：2015年天津大学生代表队征战美国第30届APEC世界Micromouse竞赛（见图1-1-4），取得了世界排名第六的好成绩。2017年至2018年，天津启诚伟业科技有限公司全额资助了在天津大学生智能鼠竞赛上获得企业命题赛冠军队，到日本东京参加第38届和第39届全日本Micromouse国际公开赛（见图1-1-5），促进学习借鉴国际智能鼠先进技术，结识众多智能鼠业界专家教授，对中国智能鼠技术的发展与提升起到推动的作用。

　　接着是蜕变升华：智能鼠大赛在中国进行本土化创新改革，设计了一系列从易到难的启诚智能鼠教学平台，满足"中、高、本、硕"不同学习阶段学生学习应用。从2016年开始先后邀请美国麻省理工学院的David Otten教授、中国台湾龙华科技大学苏景晖教授、新加坡义安理工学院黄明吉教授、英国伯明翰城市大学Peter Harrison教授、日本智能鼠国际公开赛组委会秘书长中川友纪子先生等智能鼠专家和来自泰国、印度、印尼、巴基斯坦、柬埔寨等国际"鲁班

视　频

学习借鉴
（美国
APEC）

视　频

蜕变升华
（第一届
IEEE）

工坊"师生，以及来自天津、北京、河南、河北等国内省市精英级代表队，先后加盟中国IEEE智能鼠走迷宫国际邀请赛（见图1-1-6、图1-1-7）。国际选手通过参加中国比赛，对中国竞赛标准、竞赛规则、竞赛模式和竞赛理念有了更深层次的了解和认同，从而切实推动了国际化的交流与合作，达到"互学互鉴"的目的。

图1-1-4　中国天津代表队远赴美国参加国际大赛

图1-1-5　中国天津代表队远赴日本参加国际大赛

图1-1-6　第三届IEEE智能鼠走迷宫国际邀请赛

图1-1-7 "启诚杯"第四届IEEE智能鼠走迷宫国际邀请赛

　　最后是引领辐射：教育对外开放是我国改革开放事业的重要组成部分，随着"一带一路"倡议的推进，2016年以来在中国教育部指导下，先后启动了海外鲁班工坊国际项目，智能鼠作为中国优秀的教育装备，伴随着鲁班工坊走出国门与世界分享。从2016年至今，启诚智能鼠来到泰国、印度、印尼、巴基斯坦、柬埔寨、尼日利亚、埃及等国家，免费开展智能鼠竞赛的推广和课程培训，受到了沿线国家师生的一致青睐（见图1-1-8~图1-1-13）。智能鼠成为连接世界的纽带与桥梁！

视频
引领辐射(印度鲁班)

图1-1-8 印度鲁班工坊开展智能鼠培训课程

图1-1-9 2016年泰国鲁班工坊开展智能鼠培训课程

图1-1-10　2017年印尼鲁班工坊开展智能鼠培训课程

图1-1-11　2018年巴基斯坦鲁班工坊开展智能鼠培训课程

图1-1-12　2018年柬埔寨鲁班工坊开展智能鼠培训课程

图1-1-13　2020年埃及鲁班工坊开展智能鼠培训课程

# 任务二　智能鼠的竞赛与调试环境

## 一、竞赛迷宫场地

目前，国际和国内比赛都使用同样规格的比赛场地，即一个由16×16个格子组成的方形迷宫。迷宫的"墙壁"是可以插拔的，这样就可以形成各种各样的迷宫。

如图1-1-14所示为TQD-Micromouse Maze 16×16比赛场地。迷宫底板的尺寸为2.96 m×2.96 m，上面共有 16×16 个标准迷宫单元格。图1-1-15所示为古典智能鼠迷宫挡板和立柱。

图1-1-14　TQD-Micromouse Maze 16×16
迷宫场地

图1-1-15　古典智能鼠迷宫挡板和立柱

TQD-Micromouse Maze 16×16迷宫场地规范如下：

（1）迷宫由16×16个、18 cm×18 cm大小的正方形单元所组成。

（2）迷宫的挡板高5 cm，厚1.2 cm，因此两个挡板所构成的通道的实际距离为16.8 cm，挡板将整个迷宫封闭。

（3）迷宫挡板的侧面为白色，顶部为红色。迷宫的地面为木质，颜色为哑光黑。挡板侧面和顶部的涂料能够反射红外线，地板能够吸收红外线。

（4）迷宫的起始单元可设在迷宫四个角之中的任何一个。起始单元必须三面有挡板，只留一个出口。迷宫的终点设在迷宫中央，由四个正方形单元构成。

（5）在每个单元的四角可以插上一个小立柱，其截面为正方形。如图1-1-14所示。立柱长1.2 cm、宽1.2 cm、高5 cm。小立柱所处的位置称为"格点"。除了终点区域的格点外，每个格点至少要与一面挡板相接触。

（6）迷宫制作的尺寸精度误差应不大于5%，或小于2 cm。迷宫地板的接缝不能大于0.5 mm，接合点的坡度变化不超过4°。挡板和立柱之间的空隙不大于1 mm。

（7）起点和终点设计遵照IEEE智能鼠竞赛标准，即智能鼠按照顺时针方向开始运行。

## 二、专用测试场地

专用测试场地上绘有13个标记位置，并且使用不同的颜色进行区分（见图1-1-16），用于调试红外传感器和优化转弯控制参数。接下来就带领读者认识一下它：

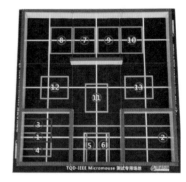

图1-1-16　TQD-IEEE Micromouse
专用测试场地

（1）①至②，灰色通道，用来检测智能鼠在无红外校准的情况下直行的偏移量。

（2）③深红色矩形，④橙色矩形；③至②、④至②均是用来检验有红外校准时的智能鼠直行情况。

（3）⑤黄色矩形用来调节智能鼠左前红外强度，⑥绿色矩形用来调节智能鼠右前红外强度；校正车姿。

（4）⑦、⑧绿色矩形用来调节智能鼠右后红外强度，⑨、⑩绿色矩形用来调节智能鼠左后红外强度，检测路口。

（5）⑪、⑫、⑬三个蓝色矩形用来调试智能鼠转弯90°。

## 三、全自动计分系统

为了精确计量智能鼠完成迷宫的时间，需要全自动地计算智能鼠通过起点和终点的时间。图1-1-17所示为由天津启诚伟业科技有限公司设计生产的用于智能鼠走迷宫竞赛的电子自动计分系统。

TQD-Micromouse Timer V2.0系统包含起点对射模块、终点对射模块、智能鼠计分系统模块、计分软件等。

图1-1-17　TQD-Micromouse Timer V2.0

起点对射模块和终点对射模块采用迷你USB充电方式，通过内置的一组激光对射传感器检测智能鼠经过。智能鼠计分系统模块用于接收起点对射模块和终点对射模块通过ZigBee发过来的数据，经过计算机中的计分软件处理，以一种直观的方式展现智能鼠在迷宫中的运行情况。计分软件也可以单独使用，可通过鼠标输入起点事件和终点事件。计分系统整体的计时精度可达0.001 s。

　　起点对射模块和终点对射模块分别安装在起点迷宫格和终点迷宫格中，如图1-1-18、图1-1-19所示。当智能鼠经过时，激光被阻断，从而产生起点或终点信号。

视频 ●

计分系统工作原理（激光的计分）

图1-1-18　迷宫起点

图1-1-19　迷宫终点

## 思考与总结

（1）IEEE国际标准智能鼠场地由哪些部分组成？

（2）请归纳智能鼠竞赛的特点。

（3）全自动计分系统大大提高了竞赛成绩计算的准确性，请简要说明其工作原理。

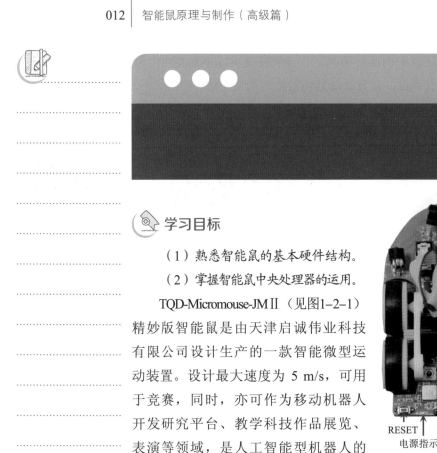

# 项目二

## 智能鼠的硬件结构

### 学习目标

（1）熟悉智能鼠的基本硬件结构。

（2）掌握智能鼠中央处理器的运用。

TQD-Micromouse-JMⅡ（见图1-2-1）精妙版智能鼠是由天津启诚伟业科技有限公司设计生产的一款智能微型运动装置。设计最大速度为 5 m/s，可用于竞赛，同时，亦可作为移动机器人开发研究平台、教学科技作品展览、表演等领域，是人工智能型机器人的典范。

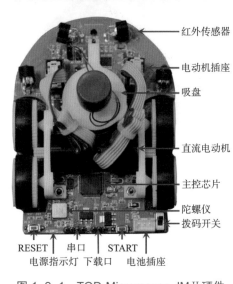

红外传感器
电动机插座
吸盘
直流电动机
主控芯片
陀螺仪
拨码开关
RESET　串口　START
电源指示灯　下载口　电池插座

图 1-2-1　TQD-Micromouse-JMⅡ硬件布局图

## 任务一　智能鼠的组成

（1）核心控制器采用高性能STM32芯片，72 MHz工作频率，512 KB的Flash存储器，64 KB的RAM，功能强大、集成度高，可高效处理智能鼠运行中的各类事件。

（2）德国原装进口高性能低功耗FAULHABER1717SR空心杯直流电动机，最大转速14 000 r/min，内置工业级一体化1024线编码器，轻松实现智能鼠速度和位置的高精度控制。

（3）ADXRS620陀螺仪检测智能鼠偏移角度，实现智能鼠运行姿态和转弯角度的高精度控制。

（4）电动机驱动采用DRV8848芯片，双路H桥设计，配合PWM技术有效响应电动机的启动、制动、正反转控制。

● 视频

智能鼠组成（JMⅡ）

（5）融入TQD吸地风扇技术，吸力可达0.5 kg，增强抓地力，有效克服轮胎打滑，从而实现高速转弯。

（6）四组模拟红外精确测距，可准确测量智能鼠离迷宫挡板的距离。

（7）动力源采用7.4 V高品质锂电池供电，优秀的稳压电路设计，并配备防反插和充电提醒功能，保证智能鼠安全稳定运行。

（8）机械结构采用高强度树脂3D打印而成，质量小，强度高，精度高，使智能鼠车身牢固，重心降低，抓地力增强，避免了高速运行时的"飞车"现象。

（9）本智能鼠采用IAR作为软件开发环境，提供开源DEMO程序包，包含红外自动检测代码和45°行走代码。

（10）中心算法和洪水算法的结合，迷宫求解更加高效，摆脱以往复杂迷宫较难搜索到终点的问题。

TQD-Micromouse-JMⅡ智能鼠的电路组成框图如图1-2-2所示，主要由核心控制器、红外发射与接收、电动机驱动、吸盘电动机、编码器与陀螺仪等组成。

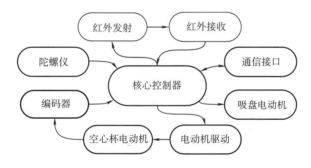

图 1-2-2 TQD-Micromouse-JMⅡ智能鼠的电路组成框图

## 任务二 智能鼠的核心控制电路

TQD-Micromouse-JMⅡ精妙版智能鼠上使用了STM32F103RET6的PWM发生器、GPIO口，SPI接口、计数器/定时器、A/D转换模块、串行口及I²C接口，其最小系统原理如图1-2-3所示。

（1）电动机驱动PWM D1_1、D1_2、D2_1、D2_2,分别对应PA10、PA11、PA8、PA9，D1_1、D1_2连接到电动机驱动芯片DRV8848的BIN1和BIN2提供左电动机驱动信号，D2_1、D2_2连接到电动机驱动芯片DRV8848的AIN1和AIN2提供右电动机驱动信号。

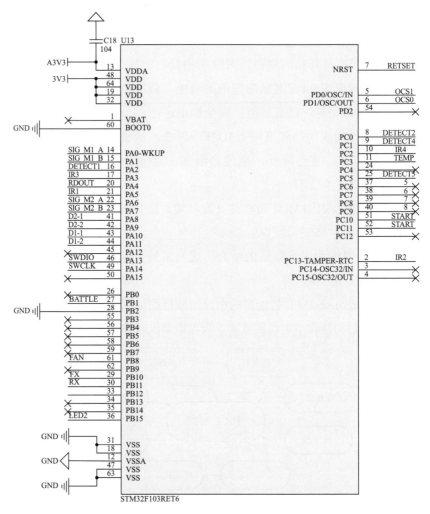

图 1-2-3 TQD-Micromouse-JM Ⅱ 最小系统原理图

（2）STLINK串行JTAG下载口SWDIO、SWCLK，分别对应PA13和PA14。

（3）编码器输出信号接口SIG_M1_A、SIG_M1_B、SIG_M2_A、SIG_M2_B，分别对应PA0、PA1、PA6、PA7，利用CPU内部的正交编码判断左右电动机正反转的编码器。

（4）四路红外发射驱动信号IR1、IR2、IR3、IR4,分别对应 PA5、PC13、PA3、PC2，提供MOS管门信号，依次对应左、右、左前、右前。

（5）四路红外接收信号 DETECT1、DETECT2、DETECT3、DETECT4，分别对应PA2、PC0、PC5、PC1，连接到红外接收头的输出端。

（6）陀螺仪接口RDOUT、TEMP,分别对应PA4、PC3,连接到ADXRS620的RDOUT和TEMP。

（7）串行口TX、RX 分别对应PB10、PB11。

（8）按键接口RESET、START,分别对应NRST、PC10和PC11。

（9）LED指示灯与PB/5连接。

## 思考与总结

（1）智能鼠的处理器是否可以由其他型号的处理器来代替？

（2）智能鼠是由传感器、控制器和执行器三部分构成的。红外线传感器相当于它的"眼睛"，可以检测四周障碍物距离，控制器依据这些信息进行处理，最后控制执行器进行动作。

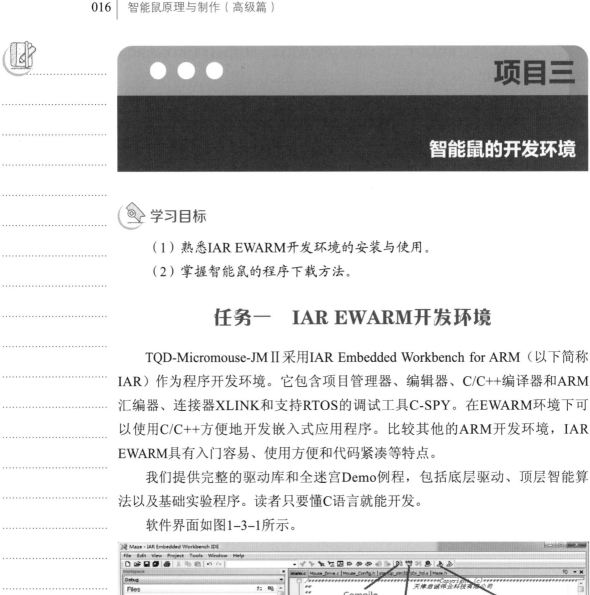

# 项目三

## 智能鼠的开发环境

### 学习目标

（1）熟悉IAR EWARM开发环境的安装与使用。

（2）掌握智能鼠的程序下载方法。

## 任务一    IAR EWARM开发环境

TQD-Micromouse-JMⅡ采用IAR Embedded Workbench for ARM（以下简称IAR）作为程序开发环境。它包含项目管理器、编辑器、C/C++编译器和ARM汇编器、连接器XLINK和支持RTOS的调试工具C-SPY。在EWARM环境下可以使用C/C++方便地开发嵌入式应用程序。比较其他的ARM开发环境，IAR EWARM具有入门容易、使用方便和代码紧凑等特点。

我们提供完整的驱动库和全迷宫Demo例程，包括底层驱动、顶层智能算法以及基础实验程序。读者只要懂C语言就能开发。

软件界面如图1-3-1所示。

● 软件

智能鼠IAR开发环境下载

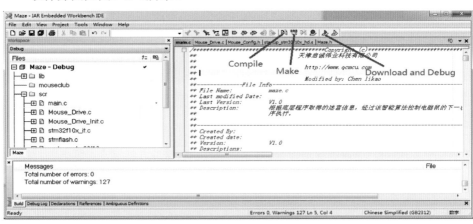

图1-3-1　IAR软件界面

# 任务二　智能鼠的程序下载

## 一、J-Link下载器

J-Link适用于单片机程序的调试与下载。该调试器结合IAR EWARM集成开发环境，可支持所有ST系列MCU的程序下载与调试。

J-Link采用USB接口与计算机连接，无论是台式计算机还是笔记本式计算机都应用自如。

软 件

J-Link驱动
程序下载

## 二、连接硬件

下载程序前一定要正确连接智能鼠、下载器和计算机。硬件连接如图1-3-2所示。

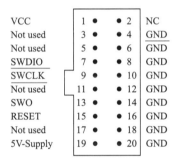
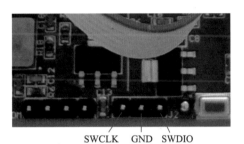

图1-3-2　硬件连接

## 三、下载程序

打开例程，双击Maze.eww打开工程，如图1-3-3所示。

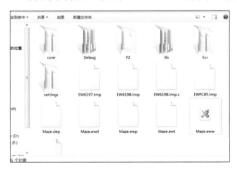

图1-3-3　打开Demo工程

（1）添加TQD库文件。在下载程序前需要添加TQD库文件，否则会出现错误。

右击工程，选择Options命令，进行设置。选择Linker选项，进入Library选项卡，如图1-3-4所示。

图1-3-4　添加TQD库文件

单击图1-3-4中的红圈图标，添加库文件。库文件位置为Demo\Debug\Exe，选择Maze.a，单击"确定"按钮。

（2）编译与下载。正确添加TQD库文件后，再进行编译与下载，就不会报错了，如图1-3-5所示。

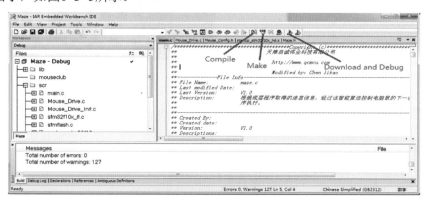

图1-3-5　下载程序

## 思考与总结

（1）常见的C语言开发软件还有哪些？

（2）IAR Embedded Workbench提供了强大的配置功能，在下载程序的时候需要根据实际情况选择下载器型号以及下载方式。

# 项目四

## 智能鼠的基本功能操作

### 学习目标

（1）了解智能鼠的红外检测原理。

（2）掌握智能鼠电动机驱动的方法。

（3）掌握智能鼠姿态检测的方法。

## 任务一 智能鼠的红外检测

传感器在控制中起了非常重要的作用，是感知系统中必不可少的部件。TQD-Micromouse-JMⅡ上共有四组模拟红外线传感器，每组传感器由红外线发射头和红外线接收头组成，如图1-4-1所示。

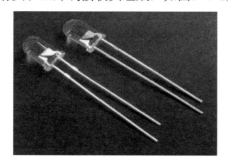

图1-4-1 TQD-Micromouse-JMⅡ红外线发射头和红外线接收头

### 一、智能鼠红外传感器收发电路

智能鼠红外传感器收发电路如图1-4-2所示。IR1~IR4为红外线发射信号驱动对应红外发射，处理器发出的红外控制信号经MOS管放大后控制红外发射管的通断，使红外发射管的发射频率保持在1 kHz，DETECT1~DETECT4为红外传感器接收到的红外反射信号，分别接入处理器的四个A/D通道，处理器则根据接收到的电压值计算当前智能鼠距障碍物的距离。

四个方向的传感器电路原理相同，红外线发射头直接由GPIO控制是否接通，按照顺序发射，从而达到四组传感器互不干扰的目的。

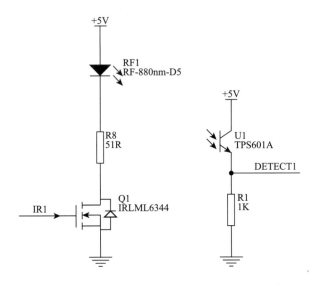

图1-4-2　TQD-Micromouse-JMⅡ红外电路[①]

## 二、红外传感器检测调试

智能鼠红外传感器用于迷宫挡板的检测，分为左前（0号）、右前（1号）、左斜（2号）、右斜（3号）四个方向，如图1-4-3所示。

图1-4-3　红外分组

按照发射方向可以分为两类：

前向类：

| 0号<br>1号 | 0号和1号检测前方，发射近、远两种不同强度的红外线。<br>作用：180°转弯（近距）、辅助90°转弯（远距）。 |
| --- | --- |

斜向类：

| 2号<br>3号 | 2号和3号检测两侧，发射三种不同强度的红外线。<br>作用：近距校正车姿、中距辅助校正车姿、远距检测路口。 |
| --- | --- |

---

① 类似图稿为Protel 99 SE导出的原理图，其图形符号与国家标准符号不一致，二者对照关系参见附录E。

TQD-Micromouse-JMⅡ智能鼠采用在线调试的方法，除红外线发射强度外，还需要进行智能鼠车姿的标定。两侧挡板距离的标定有助于智能鼠在转弯时修正车姿偏移。

　　（1）距离标定，如图1-4-4所示。

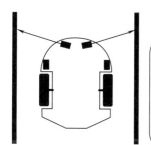

步骤一：智能鼠放在单个通道正中间。
步骤二：下载程序，单击Go按钮运行程序。
步骤三：将leftdis和rightdis的数值标定为left_distance与right_distance的数值。如图1-4-5所示的55和71（整数）。

图1-4-4　距离标定

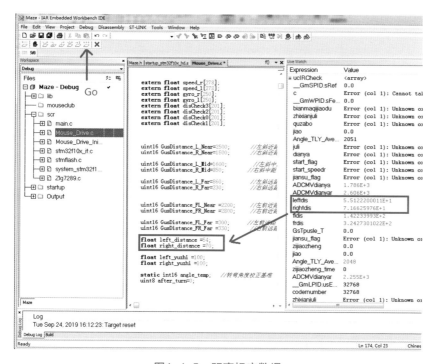

图1-4-5　距离标定数据

　　（2）0号～3号，四个红外检测强度标定：

　　①前方红外标定（0号和1号）：

　　前方远距标定如图1-4-6所示。

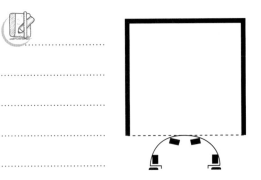

步骤一：智能鼠放在距离前方挡板一个单元格距离处。

步骤二：保持程序在运行中。

步骤三：将ucIRCheck栏中的0号和1号的数值分别标定为GusDistance_FL_Far和GusDistance_FR_Far的数值，如图1-4-7所示的753和515。

图1-4-6　前方远距标定

图1-4-7　前方远距数据

前方近距标定如图1-4-8所示。

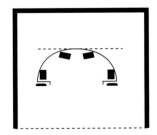

步骤一：智能鼠放在距离前方挡板3 cm处。

步骤二：保持程序在运行中。

步骤三：将ucIRCheck中的0号和1号的数值分别标定为GusDistance_FL_Near和GusDistance_FR_Near的数值，如图1-4-9所示的3852和2654。

图 1-4-8　前方近距标定

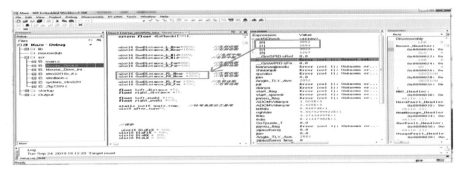

图1-4-9　前方近距数据

②斜向红外标定（2号和3号）：

左斜右斜近距标定如图1-4-10所示。

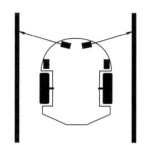

步骤一：将智能鼠放在单个通道正中间。
步骤二：保持程序在运行中。
步骤三：将ucIRCheck栏中的2号和3号的数值分别标定为GusDistance_L_Near和GusDistance_R_Near的数值，如图1-4-11所示的3235和2322。

图1-4-10 左斜右斜近距标定

图1-4-11 左斜右斜近距数据

左斜右斜远距标定如图1-4-12所示。

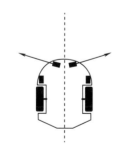

步骤一：将智能鼠放在两个通道正中间。
步骤二：保持程序在运行中。
步骤三：将ucIRCheck栏中的2号和3号的数值分别标定为GusDistance_L_Far、GusDistance_R_Far的数值，如图1-4-13所示的1025和411。

图1-4-12 左斜右斜远距标定

中距：将近距与远距（GusDistance_L_Mid和GusDistance_R_Mid）之和除以2，标定为中距。

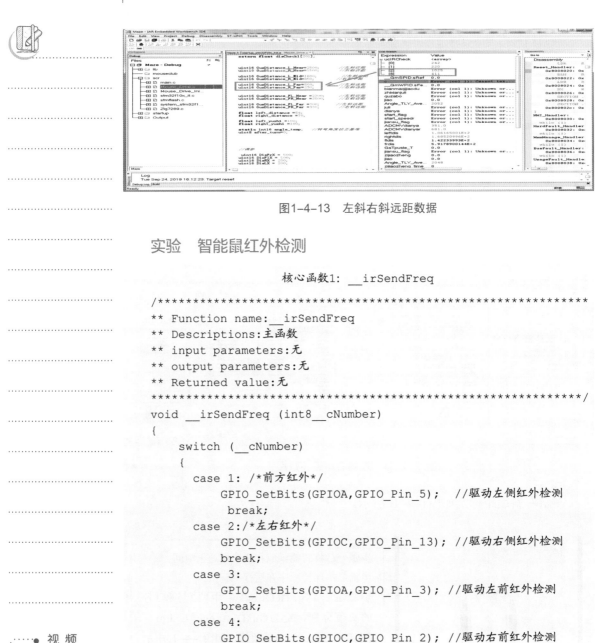

图1-4-13　左斜右斜远距数据

## 实验　智能鼠红外检测

核心函数1: __irSendFreq

```
/*******************************************************
** Function name:__irSendFreq
** Descriptions:主函数
** input parameters:无
** output parameters:无
** Returned value:无
*******************************************************/
void __irSendFreq (int8 __cNumber)
{
    switch (__cNumber)
    {
      case 1: /*前方红外*/
          GPIO_SetBits(GPIOA,GPIO_Pin_5);   //驱动左侧红外检测
           break;
      case 2:/*左右红外*/
          GPIO_SetBits(GPIOC,GPIO_Pin_13); //驱动右侧红外检测
          break;
      case 3:
          GPIO_SetBits(GPIOA,GPIO_Pin_3);  //驱动左前红外检测
          break;
      case 4:
          GPIO_SetBits(GPIOC,GPIO_Pin_2);  //驱动右前红外检测
          break;
      default:
          break;
    }
}
```

● 视频

智能鼠红外
检测

　　流程图。本实验通过驱动四组红外传感器检测周围挡板信息，在线观察返回值的变化。流程图如图1-4-14所示。

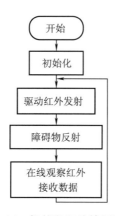

图1-4-14　智能鼠红外检测流程图

主程序:

```
                    main
/***************************************************
** Function name:main
** Descriptions:主程序
** input parameters:无
** output parameters:无
** Returned value:无
***************************************************/
main (void)
{
    uint8 n=0;                    /*有多个支路的坐标的数量*/
    uint8 ucRoadStat=0;           /*统计某一坐标可前进的支路数*/
    uint8 ucTemp=0;               /*用于START状态中坐标转换*/
    uint8 start=0;
    uint8 start_maxspeed=0;
    uint8 start_led=0;
    SystemInit();
    RCC_Init();
    JTAG_Set(1);
    MouseInit();
    :
    :
            while (1)
            {
                if (startCheck()==true)
                {
                    break;
                }
            }
        break;
```

```
        default:
            break;
        }
    }
}
```

下载程序后在线观察四组红外传感器的返回值。

# 任务二　智能鼠的电动机驱动

## 一、直流电动机

TQD-Micromouse-JMⅡ直流电动机选用FAULHABER1717SR 空心杯直流电动机（见图1-4-15），输入电压为7.4 V，可以输出1.96 W的功率，最大转速为14 000 r/min，其主要参数见表1-4-1，尺寸规格如图1-4-16所示。

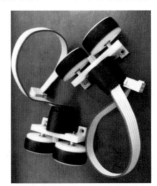

图1-4-15　FAULHABER1717SR 空心杯直流电动机

表1-4-1　空心杯直流电动机主要参数

| 序号 | 在22℃和额定电压下的值 | 003 SR | 006 SR | 012 SR | 018 SR | 024 SR | 单位 |
|---|---|---|---|---|---|---|---|
| 1 | 额定电压$U_N$ | 3 | 6 | 12 | 18 | 24 | V |
| 2 | 电枢电阻$R$ | 1.07 | 4.3 | 17.1 | 50.1 | 68.8 | Ω |
| 3 | 输出功率$P_2$ | 1.97 | 1.96 | 1.97 | 1.5 | 1.96 | W |
| 4 | 最大效率$\eta_{max.}$ | 69 | 69 | 70 | 68 | 70 | % |
| 5 | 空载转速$n_0$ | 14 000 | 14 000 | 14 000 | 12 300 | 14 000 | r/min |
| 6 | 空载电流（输出轴直径1.5 mm）$I_0$ | 0.091 | 0.046 | 0.023 | 0.013 | 0.011 | A |
| 7 | 堵转转矩$M_H$ | 5.37 | 5.34 | 5.38 | 4.66 | 5.36 | mN·m |
| 8 | 摩擦转矩$M_R$ | 0.18 | 0.18 | 0.18 | 0.18 | 0.17 | mN·m |

● 视 频

电动机扩展
知识介绍

续表

| 序号 | 在22℃和额定电压下的值 | 003 SR | 006 SR | 012 SR | 018 SR | 024 SR | 单位 |
|---|---|---|---|---|---|---|---|
| 9 | 转速常数$k_n$ | 4 820 | 2 410 | 1 210 | 709 | 602 | (r/min)/V |
| 10 | 反电动势常数$k_E$ | 0.207 | 0.414 | 0.829 | 1.41 | 1.66 | mV/(r/min) |
| 11 | 转矩常数$k_M$ | 1.98 | 3.96 | 7.92 | 13.5 | 15.9 | mN·m/A |
| 12 | 电流常数$k_I$ | 0.505 | 0.253 | 0.126 | 0.074 | 0.63 | A/(mN·m) |
| 13 | 转速/转矩斜率$\Delta n/\Delta M$ | 2 610 | 2 620 | 2 600 | 2 640 | 2 610 | r/min/ (mN·m) |
| 14 | 转子电感$L$ | 17 | 65 | 260 | 760 | 1 040 | μH |
| 15 | 机械时间常数$\tau_m$ | 16 | 16 | 16 | 16 | 16 | ms |
| 16 | 转子转动惯量$J$ | 0.59 | 0.58 | 0.59 | 0.58 | 0.59 | g·cm² |
| 17 | 最大角加速度$\alpha_{max}$ | 92 | 92 | 92 | 80 | 92 | $10^3$rad/s² |
| 18 | 热阻$R_{th1}/R_{th1}$ | 4.5/27 | | | | | kΩ/W |
| 19 | 热时间常数$\tau_{w1}/\tau_{w2}$ | 2/210 | | | | | s |
| 20 | 工作温度范围: | | | | | | |
| | — 电动机 | −30···+85（选配−55···+125） | | | | | ℃ |
| | — 线圈最高允许温度 | +125 | | | | | ℃ |
| 21 | 输出轴轴承 | 烧结轴承 | | 滚珠轴承 | 滚珠轴承预加载 | | |
| 22 | 输出轴最大载荷: | 标准 | | （选配） | （选配） | | |
| | — 输出轴直径 | 1.5 | | 1.5 | 1.5 | | mm |
| | — 3 000 r/min时，径向（距轴承3 mm） | 1.2 | | 5 | 5 | | N |
| | — 3 000 r/min时，轴向 | 0.2 | | 0.5 | 0.5 | | N |
| | — 静止,轴向 | 20 | | 10 | 10 | | N |
| 23 | 输出轴间隙 | | | | | | |
| | — 径向≤ | 0.03 | | 0.015 | 0.015 | | mm |
| | — 轴向≤ | 0.2 | | 0.2 | 0 | | mm |
| 24 | 外壳材质 | 钢，黑色涂层 | | | | | |
| 25 | 质量 | 18 | | | | | g |
| 26 | 旋转方向 | 顺时针，从正面看 | | | | | |
| 27 | 转速最大值$n_{max}$ | | | | | | r/min |
| 28 | 磁极对数 | | | | | | |
| 29 | 磁钢材料 | | | | | | |

电动机接线端与安装孔的相对角度不固定。

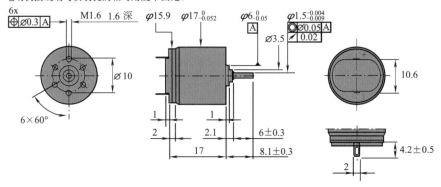

图1-4-16　空心杯直流电动机尺寸规格

## 二、驱动芯片

电动机驱动采用DRV8848芯片，电路图如图1-4-17所示。

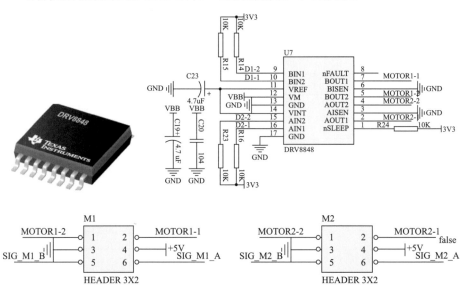

图1-4-17　电动机驱动原理图

DRV8848共包含两个H桥驱动器，每个都包含配置为全H桥的N通道和P通道功率MOSFET，用于驱动电动机绕组。每个H桥都含有一个调节电路，可通过固定关断时间斩波方案调节绕组电流。DRV8848 能够从每个H桥输出高达2 A的电流，在并联模式下驱动电流高达4 A （正常散热，12V 且TA = 25℃时）。能够很好地实现电动机启动、制动、正转及反转控制。DRV8848真值表见表1-4-2。

表1-4-2　DRV8848真值表

| PWM_H | PWM_L | 状　态 |
|---|---|---|
| L | L | 正转制动 |
| L | H | 正转 |
| H | L | 反转 |
| H | H | 反转制动 |

四种H桥驱动模式如图1-4-18所示。

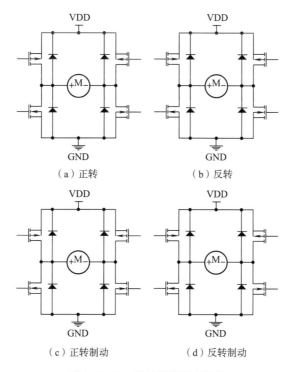

（a）正转　　　　　　　　（b）反转

（c）正转制动　　　　　　（d）反转制动

图1-4-18　四种H桥驱动模式

### 三、PWM驱动调速

图1-4-19是基本 PWM（脉宽调制）驱动电动机的电路。当开关管 V1 的栅极（$U_i$ 输入端）输入高电平时，V1 导通，电动机电枢绕组两端电压为 $U_s$。在 $t_1$ 后，栅极输入变为低电平，V1 截止，电动机的自感电流通过续流二极管 D1 迅速释放掉，电枢两端电压变为 0。$t_2$ 后，栅极输入重新变为高电平，V1 重复前面的过程。由此得到电动机电枢绕组两端的平均电压 $U_0$ 为

$$U_0=(t_1 \cdot U_s)/(t_1+t_2)=U_s \cdot (t_1/T)=\alpha \cdot U_s$$

式中，$\alpha$是占空比。

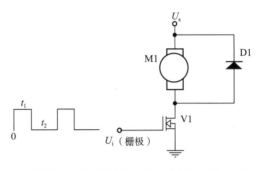

图1-4-19　基本PWM驱动电动机的电路

改变电枢电压是直流调速的主要方法。TQD-Micromouse-JMⅡ采用PWM调速方式，通过调节微处理器PWM触发信号的占空比来改变外施的平均电压$U_d$，从而实现直流电动机的调速。选用单极性PWM信号驱动直流电动机。其控制电动机正反转PWM波形如图1-4-20所示，通过改变PWM占空比调节电动机的速度大小。

● 视频

实验 智能鼠
动起来

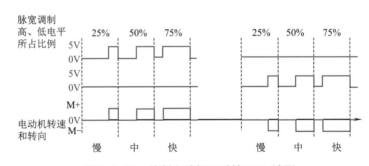

图1-4-20　控制电动机正反转PWM波形

注：M+表示电动机正转；M-表示电动机反转。

## 实验　智能鼠动起来

通过上面的学习，已经知道了电动机有多种参数需要考虑：

（1）电动机状态：启动还是停止？

（2）运行方向：向前还是向后？

（3）速度：快还是慢？

（4）需要转动的步数。

（5）已经转动的步数。

因此，可以建立一个函数结构体来存储这些参数。

结构体和其他基础数据类型一样（例如int类型、char类型），只不过结构体可以根据需要进行自定义。结构体在函数中的作用是封装。封装的好处就是可以再次利用。让使用者不必关心结构体具体是什么，只要根据定义使用就可以了。

结构体定义：电动机驱动。

```
/******************************************************
    常量宏定义——电动机加减速
******************************************************/
#define __SPEEDUP        0          /*电动机加速*/
#define __SPEEDDOWN      1          /*电动机减速*/
/******************************************************
    常量宏定义——电动机状态
******************************************************/
#define __MOTORSTOP      0          /*电动机停止*/
#define __WAITONESTEP    1          /*电动机暂停一步*/
#define __MOTORRUN       2          /*电动机运行*/
/******************************************************
    常量宏定义——电动机运行方向
******************************************************/
#define __MOTORGOAHEAD   0          /*电动机前进*/
#define __MOTORGOBACK    1          /*电动机后退*/
#define __MOTORGOSTOP    2          /*电动机反向制动*/
/******************************************************
    结构体定义
******************************************************/
struct __motor {
    int8    cState;             /*电动机运行状态*/
    int8    cDir;               /*电动机运行方向*/
    int8    cRealDir;           /*电动机运行方向*/
    uint32  uiPulse;            /*电动机需要运行的脉冲*/
    uint32  uiPulseCtr;         /*电动机已运行的脉冲*/
    int16   sSpeed;             /*当前占空比*/
};
typedef struct __motor __MOTOR;
                 核心函数1: __rightMotorContr

/******************************************************
** Function name:__rightMotorContr
** Descriptions:右直流电动机驱动
** input parameters:__GmRight.cDir（电动机运行方向）
** output parameters:无
** Returned value:无
******************************************************/
void __rightMotorContr(int speed)
{
```

```
        switch (__GmRight.cDir)
        {
        case__MOTORGOAHEAD:
          TIM_SetCompare1(TIM1,speed);
          TIM_SetCompare2(TIM1,0);
            break;

        case__MOTORGOBACK:
          TIM_SetCompare1(TIM1,0);
          TIM_SetCompare2(TIM1,speed);
            break;

        case__MOTORGOSTOP:
          TIM_SetCompare1(TIM1,0);
          TIM_SetCompare2(TIM1,0);
            break;

        default:
            break;
        }
    }
```

核心函数2: __leftMotorContr

```
/***************************************************************
** Function name:__leftMotorContr
** Descriptions:左直流电动机驱动
** input parameters:__GmLeft.cDir（电动机运行方向）
** output parameters:无
** Returned value:无
***************************************************************/
void__leftMotorContr(int speed)
{
  switch (__GmLeft.cDir)
  {
    case__MOTORGOAHEAD:
      TIM_SetCompare4(TIM1,0);
      TIM_SetCompare3(TIM1,speed);
        break;

    case__MOTORGOBACK:
      TIM_SetCompare4(TIM1,speed);
      TIM_SetCompare3(TIM1,0);
        break;

    case__MOTORGOSTOP:
      TIM_SetCompare3(TIM1,0);
      TIM_SetCompare4(TIM1,0);
        break;
```

```
default:
    break;
}
}
```

流程图。本实验学习直流电动机的驱动，通过按键次数的不同，控制智能鼠同速和差速运行。流程图如图1-4-21所示。

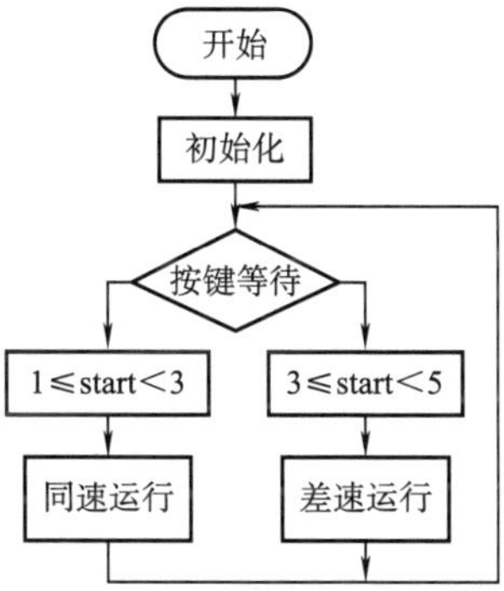

```
开始
  ↓
初始化
  ↓
按键等待 ◇
 ↓        ↓
1≤start<3   3≤start<5
 ↓        ↓
同速运行   差速运行
```

图1-4-21　按键控制智能鼠同速和差速运行

主程序：

```
                           main
/******************************************************
** Function name:main
** Descriptions:电动机驱动
** input parameters:无
** output parameters:无
** Returned value:无
******************************************************/
main (void)
{
    uint8 n=0;                  /*有多个支路的坐标的数量*/
    uint8 ucRoadStat=0;         /*统计某一坐标可前进的支路数*/
    uint8 ucTemp=0;             /*用于START状态中坐标转换*/
    uint8 start=0;
    uint8 start_maxspeed=0;
    uint8 start_led=0;
    SystemInit();
    RCC_Init();
    JTAG_Set(1);
    MouseInit();
    PIDInit();
```

```
ZLG7289Init();
delay(100000);
delay(100000);
GPIO_Config1();
USART1_Config1();
NVIC_Config1();
floodwall();
GPIO_SetBits(GPIOB,GPIO_Pin_12);
while (1) {
    if(startCheck()==true)
    {
        start=1;
    }
    if((start==1)&&GucGoHead)
    {
        __rightMotorContr(200);
        __leftMotorContr(200);
    }
    if((start==1)&&GucGoHead1)
    {
        __rightMotorContr(400);
        __leftMotorContr(0);
    }
}
}
```

# 任务三　智能鼠的姿态检测

　　由于空心杯直流电动机转速非常快，单纯依靠差速运行的时间或者脉冲数，智能鼠的转弯角度都无法达到精确控制的目的。因此，在本任务中使用陀螺仪来控制智能鼠的转弯角度。

　　TQD-Micromouse-JMⅡ采用ADXRS620陀螺仪检测旋转角度，如图1-4-22所示。ADXRS620是一种线性Z轴角速度传感器，其输出电压与角速度成正比。通过对角速度积分，便能得到角度值，从而精确控制转弯角度。

　　陀螺仪ADXRS620的角速度参考轴如图1-4-23所示，RATEOUT的输出电压与角速度成正比，当顺时针旋转时，参考轴的角速度为正值，反之为负值。

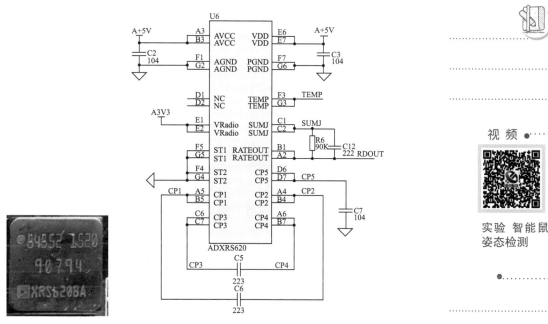

图1-4-22 ADXRS620实物图及原理图

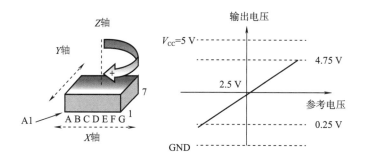

图1-4-23 陀螺仪ADXRS620的角速度参考轴

陀螺仪ADXRS620一般的零点电压为2.5 V为参考电压；输出电压与角速度的关系为6mV/(°)；当输出电压大于参考电压（2.5 V）时，角速度为正值（顺时针），反之为负值。假设$v$为陀螺仪的旋转速度，单位是(°)/s那么陀螺仪的输出电压与旋转速度$v$的关系式为

$$RATEOUT=2.5+0.006v$$

假设$v$=300(°)/s，那么

$$RATEOUT=(2.5+0.006×300) V=4.3 V$$

假设$v$=−300(°)/s,那么

$$RATEOUT=(2.5-0.006×300) V= 0.7 V$$

角速度的计算：

$$v = 1\,000\,(\text{RATEOUT}-2.5)/6$$

$$\frac{v}{\omega} = \frac{360}{2\pi} \rightarrow \omega = 2\pi v/360 \text{ rad/s}$$

角度的计算

$$\theta = \int_0^t \omega \mathrm{d}t = \int_0^t 2\pi v/360 \mathrm{d}t$$

程序设计相关计算方法：采用十位的ADJI将0～5V的陀螺仪电压分成1 023份（$2^{10}$=1 024，然后再减1）那么，2.5 V对应的A/D转换后的数字电压为511，那么

$$\omega = 2\pi \times 1\,000(\text{RATEOUT}-511)/6/360$$

$$\omega = \frac{25 \times \pi}{27} \times (\text{RATEOUT}-511)$$

$$\theta = \int_0^t w\mathrm{d}t = \sum w(i)(t)i$$

核心函数1: voltageDetect

```
/***********************************************************
** Function name:voltageDetect
** Descriptions:陀螺仪电压采集
** input parameters:无
** output parameters:无
** Returned value:无
***********************************************************/
void voltageDetect(void)
{
  u16 w;
  if(Angle_TLY_Average>=voltageDetectRef)
  {
     w=Angle_TLY_Average-voltageDetectRef;
  }
  else
  {
     w=voltageDetectRef-Angle_TLY_Average;
  }
  GW=GW+w;
}
```

流程图。本实验通过人为转动智能鼠的方式，在线观察陀螺仪参数的变化。流程图如图1-4-24所示。

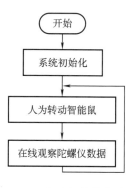

图1-4-24 观察陀螺仪输出值流程图

主程序：

main.c

```
/************************************************************
** Function name:main
** Descriptions:陀螺仪电压采集和分析
** input parameters:无
** output parameters:无
** Returned value:无
*************************************************************/
main (void)
{
    uint8 n=0;                  /*有多个支路的坐标的数量*/
    uint8 ucRoadStat=0;         /*统计某一坐标可前进的支路数*/
    uint8 ucTemp=0;             /*用于START状态中坐标转换*/
    uint8 start=0;
    uint8 start_maxspeed=0;
    uint8 start_led=0;
    SystemInit();
    RCC_Init();
    JTAG_Set(1);
    MouseInit();
    :
              while(1)
            {
                if(startCheck()==true)
                {
                    break;
                }
            }
```

```
                    break;
            default:
                break;
            }
        }
    }
```

下载程序后在线观察陀螺仪的返回值。

# 第二篇　综合实践篇

　　本篇主要介绍智能鼠直线运动和精确转弯的调试方法，并对智能鼠的高级功能进行讲解。

# 项目一

## 智能鼠的运动姿态控制

### 学习目标

（1）掌握智能鼠速度调节的原理和控制方法。

（2）掌握智能鼠精确转弯的控制方法。

智能鼠在迷宫中的运行，可以简化为直行和转弯两部分。直行是指智能鼠在迷宫中通过检测两侧挡板，校正车姿，避免碰触；转弯是指智能鼠精确转弯180°和90°，结合直行最终到达目的地。

## 任务一　智能鼠车速的闭环控制

智能鼠要完成迷宫搜索任务，首先必须具备稳定的直行速度。本书涉及的智能鼠采用PWM脉宽调制的方法来调节电动机的转速，使用单片机计数器来记录单位时间内电动机脉冲数来测量电动机的转速。

然而由于系统存在不稳定性，比如电池电压的衰减、环境温度的漂移、器件性能的变化等因素，都会造成智能鼠无法获取稳定的速度。为解决此问题，可以采用一种负反馈的闭环控制算法来控制电动机的转速。

PID控制算法（又称PID控制器）是一种常见的电动机转速控制算法。PID（比例–积分–微分）控制器作为最早实用化的控制器已有70多年历史，现在仍然是应用最广泛的工业控制器。PID控制器简单易懂，使用中不需要精确的系统模型等先决条件，因而成为应用最为广泛的控制器。

### 一、PID控制原理

PID是比例（proportion）、积分（integral）、微分（differential）的首字母缩写，分别代表了三种控制算法。通过这三种控制算法的组合可有效地纠正被控制对象的偏差，从而使其达到一个稳定的状态。

常规的模拟PID控制系统原理图如图2-1-1所示。该系统由模拟PID控制器和被控对象组成。$t$代表某一个时刻，$r(t)$是给定值，$y(t)$是系统的实际输出值，

给定值与实际输出值构成控制偏差$e(t)$，所以

$$e(t)=r(t)-y(t) \tag{2-1-1}$$

$e(t)$作为PID控制器的输入，$u(t)$作为PID控制器的输出和被控对象的输入。故模拟PID控制器的控制规律为

$$u(t)=K_{\mathrm{p}}\left[e(t)+\frac{1}{T_{\mathrm{i}}}\int_0^t e(t)\mathrm{d}t+T_{\mathrm{d}}\frac{\mathrm{d}e(t)}{\mathrm{d}t}\right] \tag{2-1-2}$$

式中，$K_{\mathrm{p}}$为控制器的比例系数；$T_{\mathrm{i}}$为控制器的积分时间，又称积分系数；$T_{\mathrm{d}}$为控制器的微分时间，又称微分系数。

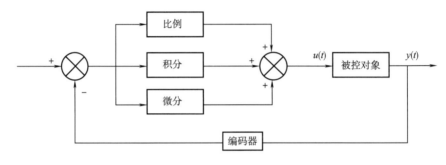

图2-1-1　模拟PID控制系统原理图

## 1. 比例部分（P）

比例部分主要是迅速对控制系统的偏差信号通过比例反应做出控制，减轻偏差信号的变化。$K_{\mathrm{p}}e(t)$是比例部分的数学表达式，需要合理地调整$K_{\mathrm{p}}$的数值，数值越大，会使系统越不稳定；数值越小，系统的调节速度越慢。比例调节的缺点是将产生静态偏差，也就是当达到稳定状态时，被控制量将产生一个固定的上下波动，如图2-1-2所示。

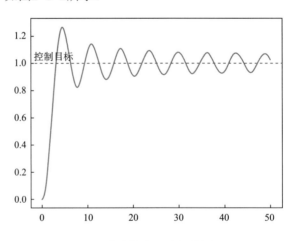

图2-1-2　产生静态误差的效果

## 2. 积分部分（I）

积分部分起到消除静态偏差的作用，$K_\text{p} \dfrac{1}{T_\text{i}} \displaystyle\int_0^t e(t)\text{d}t$ 是积分部分的数学表达式。从表达式中可以知道，只要有偏差存在，积分就会起到控制作用，直到偏差为零，如图2-1-3所示。时间常数$T_\text{i}$对积分部分有着重要作用，$T_\text{i}$的数值增大时，积分起到的作用变小，系统偏差去除的时间变长；$T_\text{i}$的数值减小时，积分起到的作用变大，系统偏差去除的时间缩短。

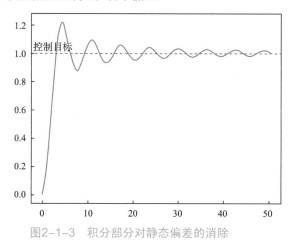

图2-1-3　积分部分对静态偏差的消除

## 3. 微分部分（D）

微分部分的作用是根据偏差的变化趋势预先给出适当的纠正。微分部分的引入，将有助于减小超调量、克服振荡，使系统趋于稳定，加快了系统的跟踪速度。微分部分的作用如图2-1-4所示。微分部分的作用对输入信号的噪声很敏感，对噪声较大的系统一般不使用微分。

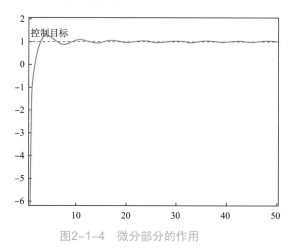

图2-1-4　微分部分的作用

## 二、增量式PID算法

若要将PID算法应用在计算机领域，就需要对PID的控制公式进行适当的变换，实现数字式的PID控制。增量式PID算法就是一种典型的数字式PID算法。

所谓增量式PID算法是指数字控制器输出的是控制量的增量 $\Delta u_k$，其中 $k$ 为离散化的时刻，一般指第 $k$ 个采样时刻。增量式PID算法可以通过式（2-1-2）推导得出。首先，可以根据式（2-1-2）得到一个离散化的表达式：

$$u_k = K_p \left[ e_k + \frac{1}{T_i} \sum_{j=0}^{k} e_j + T_d \frac{e_k - e_{k-1}(t)}{T} \right] \tag{2-1-3}$$

式中，$j$ 为计算0到 $k$ 时刻积分值时使用的循环变量；$u_k$ 为 $k$ 次采样时刻输出值；$e_k$ 为 $k$ 次采样时刻误差；$e_{k-1}$ 为 $k-1$ 次采样时刻误差；$T$ 为采样周期。

根据式（2-1-3）可以求出 $k$ 时刻与 $k-1$ 时刻输出量之差 $\Delta u_k$，如式（2-1-4）所示：

$$
\begin{aligned}
\Delta u_k = u_k - u_{k-1} &= K_p \left[ e_k - e_{k-1} + \frac{1}{T_i} e_k + T_d \frac{e_k - 2e_{k-1} + e_{k-2}}{T} \right] \\
&= K_p \left( 1 + \frac{T}{T_i} + \frac{T_d}{T} \right) e_k - K_p \left( 1 + \frac{2T_d}{T} \right) e_{k-1} + K_p \frac{T_d}{T} e_{k-2} \\
&= A e_k - B e_{k-1} + C e_{k-2}
\end{aligned}
\tag{2-1-4}
$$

式中，$A = K_p \left( 1 + \frac{T}{T_i} + \frac{T_d}{T} \right)$；$B = K_p \left( 1 + \frac{2T_d}{T} \right)$；$C = K_p \frac{T_d}{T}$。

由式（2-1-4）可以看出，如果采用恒定的采样周期 $T$，一旦确定了 $A$、$B$、$C$，只要使用前后三次测量的偏差值，就可求出控制量，其计算量比较小。最终可以得到：

$$u_k = \Delta u_k + u_{k-1} \tag{2-1-5}$$

● 视 频

实验 智能鼠
PID算法实现

### 实验 智能鼠PID算法实现

相关的宏定义和全局变量。

```
#define__KP 30      //宏定义比例系数
#define__KI 0.1     //积分系数
#define__KD 0       //微分系数
struct__pid          //定义适用于PID算法的结构体类型
{
    int16 usFeedBack;        //速度反馈值
    uint16 usEncoder_new;    //编码器
    uint16 usEncoder_last;   //编码器
    float sRef;
```

```
    float sFeedBack;
    float sPreError;    //速度误差, vi_Ref-vi_FeedBack
    float sPreDerror;   //速度误差之差, d_error-PreDerror;
    fp32 fKp;           //P
    fp32 fKi;           //I
    fp32 fKd;           //D
    int16 iPreU;        //电动机控制输出值
};
typedef struct__pid__PID;
__PID  __GmSPID;        //声明用于控制直线速度的PID变量
__PID  __GmWPID;        //声明用于控制转弯速度的PID变量
```

核心函数1: PIDInit

```
/*****************************************************************
** Function name:PIDInit
** Descriptions:PID初始化
** input parameters:无
** output parameters:无
** Returned value:无
*****************************************************************/
void PIDInit(void)
{
    __GmLPID.usEncoder_new=32768;
    __GmLPID.usFeedBack=0;  //速度反馈值
    __GmLPID.sFeedBack=0;
    __GmRPID.usEncoder_new=32768;
    __GmRPID.usFeedBack=0;  //速度反馈值
    __GmRPID.sFeedBack=0;
    __GmSPID.sRef=0;        //速度设定值
    __GmSPID.sFeedBack=0;
    __GmSPID.sPreError=0;   //速度误差, vi_Ref-vi_FeedBack
    __GmSPID.sPreDerror=0;  //速度误差之差, d_error-PreDerror
    __GmSPID.fKp=__KP;      //
    __GmSPID.fKi=__KI;      //
    __GmSPID.fKd=__KD;
    __GmSPID.iPreU=0;       //电动机控制输出值
    __GmWPID.sRef=0;        //速度设定值
    __GmWPID.sFeedBack=0;
    __GmWPID.sPreError=0;   //速度误差, vi_Ref-vi_FeedBack
    __GmWPID.sPreDerror=0;  //速度误差之差, d_error-PreDerror
    __GmWPID.fKp=__KP;      //经验值为30
    __GmWPID.fKi=__KI;      //经验值为0.1或0.01
    __GmWPID.fKd=__KD;
    __GmWPID.iPreU=0;       //电动机控制输出值
}
```

核心函数2:　__SPIDContr

```
/***************************************************************
** Function name:__SPIDContr
** Descriptions:直线PID控制
** input parameters:无
** output parameters:无
** Returned value:无
***************************************************************/
void__SPIDContr(void)
{
    float   error,d_error,dd_error;
    static uint8   K_I=1;
    error=__GmSPID.sRef-__GmSPID.sFeedBack; // 偏差计算
    d_error=error-__GmSPID.sPreError;
    dd_error=d_error-__GmSPID.sPreDerror;
    if(error>Deadband)
      error-=Deadband;
    else if(error<-Deadband)
      error+=Deadband;
    else
      error=0;
    if((error>error_IMAX)||(error<-error_IMAX))
      K_I=0;
    else
      K_I=1;
    __GmSPID.sPreError=error; //存储当前偏差
    __GmSPID.sPreDerror=d_error;
    __GmSPID.iPreU+=(int16)(__GmSPID.fKp*d_error+K_I*__GmSPID.
fKi*error+__GmSPID.fKd*dd_error);
    }
```

核心函数3:　__WPIDContr

```
/***************************************************************
** Function name:__WPIDContr
** Descriptions:转向PID控制
** input parameters:无
** output parameters:无
** Returned value:无
***************************************************************/
void__WPIDContr(void)
{
    float   error,d_error,dd_error;
    static uint8   K_I=1;
    error=__GmWPID.sRef+0.5*GsTpusle_T-__GmWPID.sFeedBack; // 偏差计算
    d_error=error-__GmWPID.sPreError;
    dd_error=d_error-__GmWPID.sPreDerror;
```

```
    if(error>Deadband)
      error-=Deadband;
    else if(error<-Deadband)
      error+=Deadband;
    else
      error=0;
    if((error>error_IMAX)||(error<-error_IMAX))
      K_I=0;
    else
      K_I=1;
    __GmWPID.sPreError=error; //存储当前偏差
    __GmWPID.sPreDerror=d_error;
    __GmWPID.iPreU+=(int16)(__GmWPID.fKp*d_error+K_I*__GmWPID.
fKi*error+__GmWPID.fKd*dd_error);
  }
```

核心函数4：__PIDContr

```
/***************************************************************
** Function name:__PIDContr
** Descriptions:PID控制，通过脉冲数控制电动机
** input parameters:无
** output parameters:无
** Returned value:无
***************************************************************/
void__PIDContr(void)
{
    __SPIDContr();
    __WPIDContr();
    __GmLeft.sSpeed=__GmSPID.iPreU-__GmWPID.iPreU ;
    if(__GmLeft.sSpeed>=0){
    __GmLeft.cDir=__MOTORGOAHEAD;
    if(__GmLeft.sSpeed>=U_MAX)    //最大速度调节，防止溢出
      __GmLeft.sSpeed=U_MAX;
    if(__GmLeft.sSpeed<=U_MIN)    //最小速度调节，防止溢出
      __GmLeft.sSpeed=U_MIN;
    }
    else{
      __GmLeft.cDir=__MOTORGOBACK;
      __GmLeft.sSpeed*=-1;
    if(__GmLeft.sSpeed>=U_MAX)
      __GmLeft.sSpeed=U_MAX;
    if(__GmLeft.sSpeed<=U_MIN)
      __GmLeft.sSpeed=U_MIN;
    }
    __GmRight.sSpeed=__GmSPID.iPreU+__GmWPID.iPreU;
    if(__GmRight.sSpeed>=0){
```

```
        __GmRight.cDir=__MOTORGOAHEAD;
    if(__GmRight.sSpeed>=U_MAX)
        __GmRight.sSpeed=U_MAX;
    if(__GmRight.sSpeed<=U_MIN)
        __GmRight.sSpeed=U_MIN;
    }
    else{
        __GmRight.cDir=__MOTORGOBACK;
        __GmRight.sSpeed*=-1;
    if(__GmRight.sSpeed>=U_MAX)
        __GmRight.sSpeed=U_MAX;
    if(__GmRight.sSpeed<=U_MIN)
        __GmRight.sSpeed=U_MIN;
    }
    __rightMotorContr();
    __leftMotorContr();
}
```

流程图。本实验通过修改__GmWPID.sRef和__GmSPID.sRef这两个数值来控制智能鼠以稳定的速度前进或者转弯。流程图如图2-1-5所示。

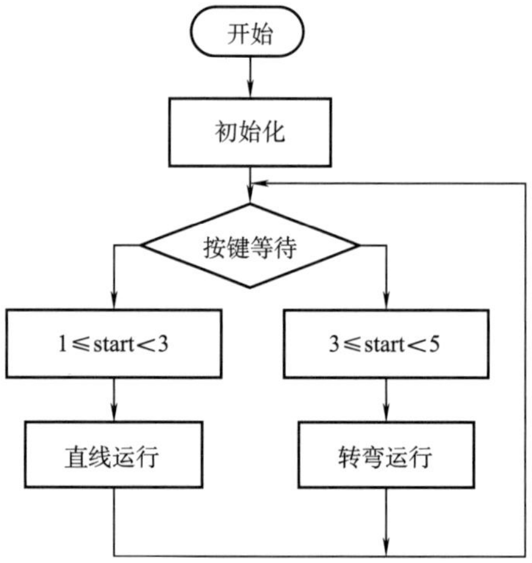

图2-1-5　智能鼠PID控制流程图

SysTick_Handler

```
void SysTick_Handler(void)
{
    __Encoder();//获取电动机脉冲值及速度值
    __PIDContr();//使用PID算法控制电动机速度
}
```

main.c

```
/***********************************************************
** Function name:main
** Descriptions:主函数
** input parameters:无
** output parameters:无
** Returned value:无
***********************************************************/
main (void)
{
    uint8 start=0;
    SystemInit();
    RCC_Init();
    JTAG_Set(1);
    MouseInit();
    PIDInit();
    ZLG7289Init();
    delay(100000);
    delay(100000);
    GPIO_Config1();
    USART1_Config1();
    NVIC_Config1();
    floodwall();
    GPIO_SetBits(GPIOB,GPIO_Pin_12);
    while (1) {
        if(startCheck()==true)
        {
            start++;
        }
        if((start<3)&&GucGoHead)
        {
            __GmSPID.sRef=100;
            __GmWPID.sRef=0;
        }
        if((start>=3)&&GucGoHead)
        {
            __GmSPID.sRef=100;
            __GmWPID.sRef=100;
        }
    }
}
```

## 任务二　智能鼠转弯的精确控制

　　智能鼠在迷宫中运行时除直行外，还会遇到大量转弯。转弯是否准确，对智能鼠的后续运行影响巨大。

智能鼠常用的转弯方式有三种，如图2-1-6所示。

第一种、第二种转弯方式控制简单，整个过程清晰明确，稳定性较高，但转弯用时较多，步进电动机通常会采用这两种方式；第三种转弯方式速度较快，但是控制难度稍大，直流电动机通常采用这种方式。

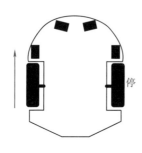
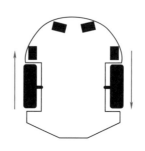
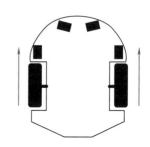

图2-1-6　三种转弯方式

## 实验　智能鼠优化转弯

智能鼠在迷宫中转弯时，除需要考虑转弯角度外，还需要考虑转弯的时机。转弯太早或太晚都会对智能鼠后续的运行产生影响。

智能鼠在$t_0$进入转弯过程，$t_0$至$t_1$是转弯前脉冲运行时间，$t_1$至$t_2$是转弯时间，$t_2$转弯结束，$t_2$至$t_3$是转弯后脉冲运行时间，如图2-1-7所示。

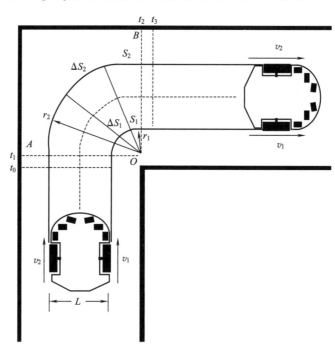

● 视频

实验 智能鼠
优化转弯

图2-1-7　智能鼠转弯整体过程

### 1. 转弯前后脉冲控制

若转弯前脉冲过大，转弯后会贴近外侧挡板；若太小，转弯后会贴近内侧挡板，如图2-1-8所示。

转弯后的脉冲大小又会对下一个转弯造成影响。

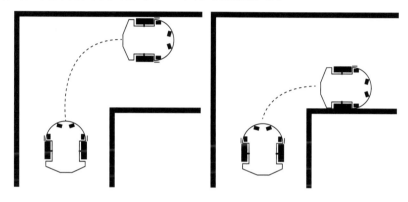

图 2-1-8 转弯前脉冲调节示意图

核心函数1: mazeSearch

```
/**********************************************************
** Function name:mazeSearch
** Descriptions:搜索程序
** input parameters:无
** output parameters:无
** Returned value:无
**********************************************************/
static int16   GuiTpusle_LR_l=0;
static int16   GuiTpusle_LR_r=0;
static uint8   sjjj;
void mazeSearch(void)
{
.
.
.
if (cL) {                              /*是否允许检测左边*/
            if((__GucDistance[__LEFT)&0x01)==0)
            {                         /*如果左边有支路，跳出程序*/
                __GmRight.uiPulse=__GmRight.uiPulseCtr+17800-
GuiTpusle_LR-GuiTpusle_LR_l;
                __GmLeft.uiPulse=__GmLeft.uiPulseCtr+17800-
GuiTpusle_LR-GuiTpusle_LR_l;
   /* 左转前脉冲*/
.
.
.
if (cR) {                              /*是否允许检测右边*/
```

```
                        if ((__GucDistance[__RIGHT]&0x01)==0){
                                          /*如果右边有支路，跳出程序*/
                        __GmRight.uiPulse=__GmRight.uiPulseCtr+19800-
GuiTpusle_LR-GuiTpusle_LR_r;
                        __GmLeft.uiPulse=__GmLeft.uiPulseCtr+19800-
GuiTpusle_LR-GuiTpusle_LR_r;
                                                  /* 右转前脉冲*/

                        :
                        :
                        }
```

核心函数2：mouseTurnright（部分程序）

```
/*************************************************************
** Function name:mouseTurnright
** Descriptions:右转
** input parameters:无
** output parameters:无
** Returned value:无
*************************************************************/
void mouseTurnright(void)
{
    :
    :
    __GmLeft.uiPulse=12000- piancha_r*2886.67;
    __GmLeft.uiPulseCtr=0;
    __GmRight.uiPulse=12000-piancha_r*2886.67;//右转后脉冲（左转后脉冲类似）
    __GmRight.uiPulseCtr=0;
    :
    :
}
```

## 2. 转弯速度控制

智能鼠在转弯时，直线速度越快，越容易打滑；旋转速度越快，转弯半径越小。所以，需要多次试验找出最完美的参数。

直线速度SPID=(V1+V2)/2。

旋转速度WPID=(V1−V2)/2。

核心函数3：mouseTurnright（全程序）

```
/*************************************************************
** Function name:mouseTurnright
** Descriptions:右转
** input parameters:无
** output parameters:无
** Returned value:无
```

```
************************************************************/
static uint8 after_turnright;
static uint8 aaaaa;
static uint8 lwt1;
float bc_r;
static int sssss=0;
void mouseTurnright(void)
{
static int ir=0;
float leftd=left_distance;
  piancha_r=pianchar(leftdis,leftd);
  GW=0;
  time=0;
  __GucMouseState=__TURNRIGHT;
  __GmRight.cState=__MOTORRUN;
  __GmLeft.cState=__MOTORRUN;
  GucMouseDir=(GucMouseDir+2)%8;              /*方向标记*/
  __GmSPID.sRef=145;                          //直线速度
  __GmWPID.sRef=-86;                          //旋转速度
   while(1)
   {
       if(GW>150000)
        {
          break;
        }
   }
   ir++;
   __GmWPID.sRef=0;
   __GucMouseState=__GOAHEAD;
   GuiSpeedCtr=3;
   __GmLeft.uiPulse=12000- piancha_r*2886.67;
   __GmLeft.uiPulseCtr=0;
   __GmRight.uiPulse=12000- piancha_r*2886.67;
   __GmRight.uiPulseCtr=0;

   while ((__GmRight.uiPulseCtr+200)<=__GmRight.uiPulse);
   while ((__GmLeft.uiPulseCtr+200)<=__GmLeft.uiPulse);
   __GucMouseState=__TURNRIGHT;
   GuiSpeedCtr=__SPEEDUP;
    __GmRight.cState=__MOTORSTOP;
    __GmLeft.cState=__MOTORSTOP;
    __GmRight.sSpeed=0;
    __rightMotorContr();
    __GmLeft.sSpeed=0;
    __leftMotorContr();
    __GmRight.uiPulseCtr=0;
    __GmLeft.uiPulseCtr=0;
```

```
mouseStop();
while(1);
}
```

流程图。本实验以直行加右转为例来验证智能鼠转弯的整体过程。流程图如图2-1-9所示。

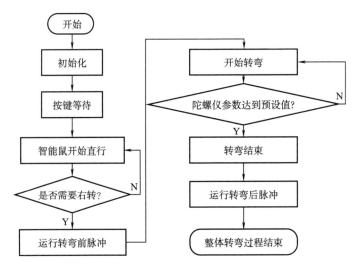

图2-1-9　智能鼠转弯整体过程验证流程图

主程序：

main.c

```
/***********************************************************
** Function name:main
** Descriptions:主函数
** input parameters:无
** output parameters:无
** Returned value:无
***********************************************************/
main (void)
{
    uint8 n=0;                  /*有多个支路的坐标的数量*/
    uint8 ucRoadStat=0;         /*统计某一坐标可前进的支路数*/
    uint8 ucTemp=0;             /*用于START状态中坐标转换*/
    uint8 start=0;
    uint8 start_maxspeed=0;
    uint8 start_led=0;
    SystemInit();
    RCC_Init();
    JTAG_Set(1);
```

```
    MouseInit();
    PIDInit();
    ZLG7289Init();
    delay(100000);
    //__ir_Get();
    delay(100000);
    GPIO_Config1();
    USART1_Config1();
    NVIC_Config1();
    floodwall();
    GPIO_SetBits(GPIOB,GPIO_Pin_12);
    while(1) {
        switch (GucMouseTask) {            /*状态机处理*/
            case WAIT:
                sensorDebug();
                delay(10000);

                if (startCheck()==true)
                    {
                        start++;
                     }
                if(start=1&&GucGoHead)
                    {
                        start=0;
                        GucMouseTask=START;
                        delay(1000000);
                    }
                    break;
            case START:

                mazeSearch();

                 while(1)
                 {
                     if (startCheck()==true)
                     {
                         break;
                     }
                 }
                    break;
            default:
                break;
        }
    }
}
```

## 思考与总结

（1）智能鼠在进行闭环控制时，P、I、D对系统各有什么影响？

（2）智能鼠转弯角度控制有哪些方法？

（3）为了实现对电动机速度和转弯角度的精确控制，我们引入了PID控制理念以及陀螺仪。PID闭环控制确保车速的稳定，陀螺仪实时检测智能鼠偏移角度，从而达到车速和姿态的高精度控制。

# 项目二

## 智能鼠的智能控制算法和技术

### 学习目标

（1）了解智能鼠校正走直线的原理。

（2）熟悉智能鼠常用的路程记录方法。

（3）掌握智能鼠的坐标获取与吸盘控制方法。

智能鼠在运行时依据传感器检测结果来校正车姿，并根据运行的路程长度与转弯方向来判断当前坐标。吸盘的应用增强了智能鼠抓地力。

## 任务一　智能鼠的智能避障

智能鼠在迷宫中的运行过程可以分为直行和转弯。准确的红外检测以及转弯参数决定了智能鼠能否顺利通过迷宫找到终点。我们已经在第一篇项目三、第二篇项目一中分别学习过红外检测以及转弯角度的精确控制。接下来就运用这两方面的知识来实现智能鼠的智能避障。

### 实验一　校正走直线

在电动机驱动任务中已经了解到，没有两个转速完全相同的电动机。为了使智能鼠能够沿着迷宫中心线行走，加入了编码器和红外线传感器。编码器可以记录电动机转速，但由于轮胎打滑等原因，需要红外线传感器来检测两侧挡板距离，判断智能鼠是否出现偏移，并针对这一情况进行校正。

核心函数1: SysTick_Handler

```
/**********************************************************
** Function name:SysTick_Handler
** Descriptions:定时中断扫描函数
** input parameters:无
** output parameters:无
** Returned value:无
```

```
***************************************************************/
void SysTick_Handler(void)
{
  float leftbiao=left_distance;       //标定值
  float rightbiao=right_distance;
  float leftyu=left_yuzhi;
  float rightyu=right_yuzhi;
  static int8 n=0,m=0,k=0,l=0,a=0,b=0,c=0,w=0;
  static u16  t=0, s=0;
  static u32  encoder=0;
  uint16 Sp;
  if(zijiaozheng0==1)
  {
        if(zijiaozheng_flag)
        {
          sum_zijiaozheng+=Angle_TLY_Average;
          zijiaozheng_time++;
          if(zijiaozheng_time==3000){zijiaozheng=sum_
zijiaozheng/3000;zijiaozheng_flag=0;jiao=0;}
        }
  }
     __Encoder();
    TIM6_IRQHandler();
    Sp=__GmSPID.sFeedBack;
    encoder+=Sp;
    t++;
    s++;
    if(t==1000)
    {
        t=0;
        GPIO_SetBits(GPIOB, GPIO_Pin_12);
        if(s==2000)
        {
            s=0;
            GPIO_ResetBits(GPIOB, GPIO_Pin_12);
        }
    }
    switch (__GmRight.cState) {
            case__MOTORSTOP:    /*停止，同时清零速度和脉冲值*/
            __GmRight.uiPulse=0;
            __GmRight.uiPulseCtr=0;
            __GmLeft.uiPulse=0;
            __GmLeft.uiPulseCtr=0;
            break;
          case__WAITONESTEP:                /*暂停一步*/
            __GmRight.cState=__MOTORRUN;
```

```
                GsTpusle_T=dis(leftdis, rightdis, leftbiao,
leftyu, rightbiao, rightyu);//
                break;
            case __MOTORRUN:                    /*电动机运行*/
              if (__GucMouseState == __GOAHEAD)
                                        /*根据传感器状态微调电动机速度*/
                {
                    //搜索冲刺直线校正
                    GsTpusle_T=disr(leftdis, rightdis, leftbiao,
leftyu, rightbiao, rightyu);//
                    if(GuiSpeedCtr==__SPEEDUP)
                    {
                      k=(k+1)%5;//20
                      if(k==4)
                      __SpeedUp();
                    }
                    else if(GuiSpeedCtr==__SPEEDDOWN)
                    {
                        k=(k+1)%10;
                        if(k==5)
                        __SpeedDown();
                    }
                    else;
                }
                else
                {
                    GsTpusle_T=0;
                    voltageDetect();          //读取陀螺仪信息
                }
                __PIDContr();
                break;
            case 4:
                GsTpusle_T=0;
                __PIDContr();
                break;
            default:
                break;
        }
    }
```

流程图。本实验使用传感器检测两侧挡板距离,来判断智能鼠是否发生偏移。流程图如图2-2-1所示。

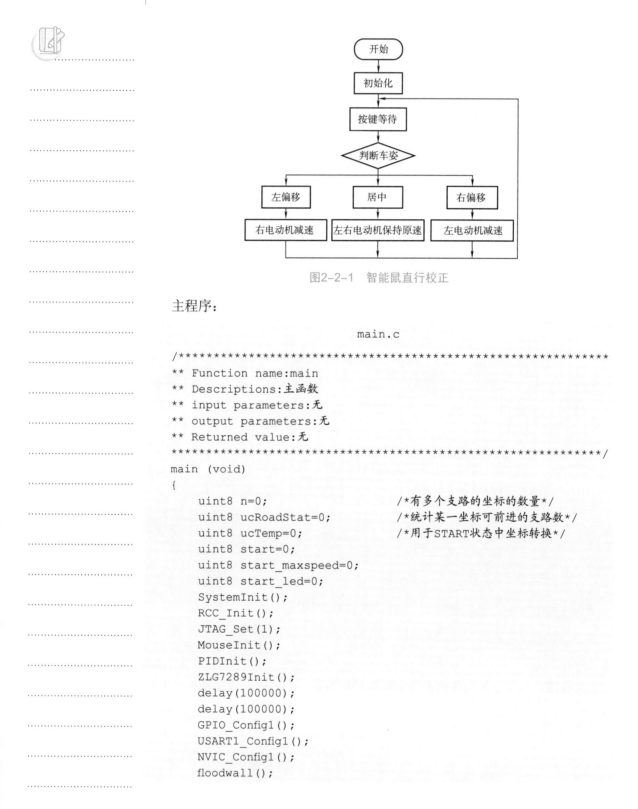

图2-2-1　智能鼠直行校正

主程序：

main.c

```
/***********************************************************
** Function name:main
** Descriptions:主函数
** input parameters:无
** output parameters:无
** Returned value:无
***********************************************************/
main (void)
{
    uint8 n=0;                      /*有多个支路的坐标的数量*/
    uint8 ucRoadStat=0;             /*统计某一坐标可前进的支路数*/
    uint8 ucTemp=0;                 /*用于START状态中坐标转换*/
    uint8 start=0;
    uint8 start_maxspeed=0;
    uint8 start_led=0;
    SystemInit();
    RCC_Init();
    JTAG_Set(1);
    MouseInit();
    PIDInit();
    ZLG7289Init();
    delay(100000);
    delay(100000);
    GPIO_Config1();
    USART1_Config1();
    NVIC_Config1();
    floodwall();
```

```
    GPIO_SetBits(GPIOB,GPIO_Pin_12);
    while (1) {
        switch (GucMouseTask) {              /*状态机处理*/
            case WAIT:
                sensorDebug();
                delay(10000);
              if(startCheck()==true)
               {
                   start++;
                }
                if(start=1&&GucGoHead)
                {
                   start=0;
                   GucMouseTask=START;
                   delay(1000000);
                }
                 break;
            case START:
            mazeSearch();
             while (1)
             {
                 if(startCheck()==true)
                 {
                     break;
                 }
             }
            break;
            default:
            break;
        }
    }
}
```

## 实验二 智能鼠自动转弯

智能鼠自动转弯的前提是红外线传感器检测到路口。再结合自身法则，挑选合适的转弯方向。

### 核心函数1: TIM6_IRQHandler

```
/**************************************************************
** Function name:TIM6_IRQHandler
** Descriptions:定时器函数
** input parameters:无
** output parameters:无
** Returned value:无
**************************************************************/
```

```
void TIM6_IRQHandler(void)
{
    :
    :

            if(ucIRCheck[2]>GusDistance_L_Far)
            {
                __GucDistance[__LEFT]|=0x01;
            }
            else
            {
                __GucDistance[__LEFT]&=0xfe;
            }

            if(ucIRCheck[2]>GusDistance_L_Mid)
            {
                __GucDistance[__LEFT]|=0x02;
            }
            else
            {
                __GucDistance[__LEFT]&=0xfd;
            }

            if(ucIRCheck[2]>GusDistance_L_Near)
            {
                __GucDistance[__LEFT]|=0x04;
            }
            else
            {
                __GucDistance[__LEFT]&=0xfb;
            }
    :
    :
```

核心函数2: mazeSearch

```
/**************************************************************
** Function name:mazeSearch
** Descriptions:搜索程序
** input parameters:无
** output parameters:无
** Returned value:无
**************************************************************/
static int16    GuiTpusle_LR_l=0;
static int16    GuiTpusle_LR_r=0;
static uint8    sjjj;
void mazeSearch(void)
{
    int8 cL=0, cR=0, cCoor=1;
```

```
if (__GmLeft.cState)
{
    cCoor=0;
}
if((__GucMouseState==__TURNRIGHT)||(__GucMouseState==__TURNLEFT))
{
    __GmLeft.uiPulseCtr=40000;
    __GmRight.uiPulseCtr=40000;
    cL=1;
    cR=1;
    if(((__GucDistance[__FRONTR]!=0)&&(__GucDistance[__
FRONTL]!=0))||((__GucDistance[__LEFT]&0x01)==0)||((__GucDistance[__
RIGHT]&0x01)==0))
    {
        if((__GucDistance[__FRONTR]!=0)&&(__GucDistance[__FRONTL]!=0))
        {
        GuiTpusle_LR =16000;
            if((__GucMouseState==__TURNRIGHT)&&((__GucDistance[__
LEFT]&0x01)==0)){GuiTpusle_LR_l=500;W_l=0;}
            if((__GucMouseState==__TURNRIGHT)&&((__GucDistance[__
RIGHT]&0x01)==0)){GuiTpusle_LR_r=0;W_r=-0;}
                if((__GucMouseState==__TURNLEFT)&&((__GucDistance[__
LEFT]&0x01)==0)){GuiTpusle_LR_l=0;W_l=0;}
                if((__GucMouseState==__TURNLEFT)&&((__GucDistance[__
RIGHT]&0x01)==0)){GuiTpusle_LR_r=-4000;W_r=0;}
        }
        else
            GuiTpusle_LR=12000;
    }
    else{
        GuiTpusle_LR=0;
    }
}
else{
    GuiTpusle_LR=0;
}
__GucMouseState=__GOAHEAD;
__GiMaxSpeed=SEARCHSPEED;
__GmRight.uiPulse=MAZETYPE*ONEBLOCK;
__GmLeft.uiPulse=MAZETYPE*ONEBLOCK;
__GmRight.cState=__MOTORRUN;
__GmLeft.cState=__MOTORRUN;
 GuiSpeedCtr=__SPEEDUP;
while (__GmLeft.cState!=__MOTORSTOP)
{
    if (__GmLeft.uiPulseCtr>=ONEBLOCK)
    {                              /*判断是否走完一格*/
```

```
                                     __GmLeft.uiPulse-=ONEBLOCK;
                                     __GmLeft.uiPulseCtr-=ONEBLOCK;
                                     if (cCoor)
                                     {
             if(((__GucDistance[__FRONTR]!=0)&&(__GucDistance[__FRONTL]!=0))&&(u
cIRCheck[2]>GusDistance_L_Far)&&(ucIRCheck[3]>GusDistance_R_Far))//0x01
                                         {
                                             GucFrontNear=1;
                                             goto End;
                                         }
                                         __mouseCoorUpdate();          /*更新坐标*/
                                     }
                                     else
                                     {
                                         cCoor=1;
                                     }
                                 }
                                 if (__GmRight.uiPulseCtr>=ONEBLOCK) {
                                                                       /*判断是否走完一格*/
                                     __GmRight.uiPulse-=ONEBLOCK;
                                     __GmRight.uiPulseCtr-=ONEBLOCK;
                                 }
                                 if (cL) {                    /*是否允许检测左边*/
                                     if ((__GucDistance[__LEFT]  & 0x01)==0)
                                     {                        /*左边有支路，跳出程序*/
                                         __GmRight.uiPulse=__GmRight.uiPulseCtr+17800-
GuiTpusle_LR-GuiTpusle_LR_1;
                                         __GmLeft.uiPulse=__GmLeft.uiPulseCtr+17800-
GuiTpusle_LR-GuiTpusle_LR_1;
                                         while ((__GucDistance[__LEFT]&0x01)==0)
                                         {
                                             if ((__GmLeft.uiPulseCtr+400)>__GmLeft.uiPulse)
                                             {
                                                 GucFrontNear=0;
                                                 goto End;
                                             }
                                         }
                                         __GmRight.uiPulse=MAZETYPE*ONEBLOCK;
                                         __GmLeft.uiPulse=MAZETYPE*ONEBLOCK;
                                         GuiSpeedCtr=__SPEEDUP;
                                     }
                                 } else {                        /*左边有挡板时开始允许检测左边*/
                                     if (ucIRCheck[2]>GusDistance_L_Far) {
                                         cL=1;
                                     }
                                 }
                                 if (cR) {                    /*是否允许检测右边*/
```

```
                  if ((__GucDistance[__RIGHT]&0x01)==0){//
mouseStop();while(1); /*右边有支路，跳出程序*/
                        __GmRight.uiPulse=__GmRight.uiPulseCtr+19800-
GuiTpusle_LR-GuiTpusle_LR_r;
                        __GmLeft.uiPulse=__GmLeft.uiPulseCtr+19800-
GuiTpusle_LR-GuiTpusle_LR_r;
                  while ((__GucDistance[__RIGHT]&0x01)==0) {
                        if((__GmLeft.uiPulseCtr+400)>__GmLeft.uiPulse)
                        {
                              GucFrontNear=0;
                              goto End;
                        }
                  }
                  __GmRight.uiPulse=MAZETYPE*ONEBLOCK;
                  __GmLeft.uiPulse=MAZETYPE*ONEBLOCK;
                  GuiSpeedCtr=__SPEEDUP;
            }
      } else {
            if (ucIRCheck[3]>GusDistance_R_Far)
            {
                  cR=1;
            }
      }
   }
End:
      __mouseCoorUpdate();                /*更新坐标*/
}


               核心函数3: crosswayChoice
/***************************************************************
** Function name:crosswayChoice
** Descriptions:选择一条支路作为前进方向
***************************************************************/
void crosswayChoice (void)
{
    switch (SEARCHMETHOD) {//依据SEARCHMETHOD判断转弯方向
    caseRIGHTMETHOD:
    mouseTurnright();           //右转弯
    break;
    caseLEFTMETHOD:
    mouseTurnleft();            //左转弯
    break;
    caseCENTRALMETHOD:
    centralMethod();
    break;
    caseFRONTRIGHTMETHOD:
```

```
            frontRightMethod();
            break;
        default:
            break;
        }
    }
```

流程图。智能鼠依据传感器的检测结果实现自动转弯。由于转弯时需要考虑的情况比较多，本实验以仅有一个路口的情况为例进行讲解。流程图如图2-2-2所示。

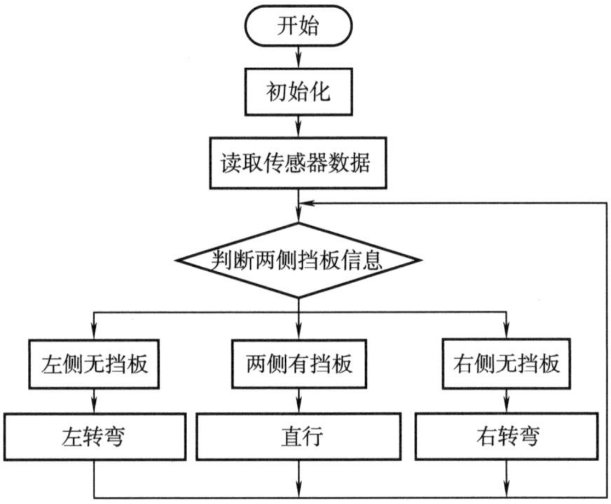

图2-2-2　智能鼠自动转弯流程图

主程序：

```
                        main.c
/***********************************************************
** Function name:main
** Descriptions:智能鼠自动转弯程序
** input parameters:无
** output parameters:无
** Returned value:无
***********************************************************/
main (void)
{
    uint8 n=0;                      /*有多个支路的坐标的数量*/
    uint8 ucRoadStat=0;             /*统计某一坐标可前进的支路数*/
    uint8 ucTemp=0;                 /*用于START状态中坐标转换  */
    uint8 start=0;
    uint8 start_maxspeed=0;
    uint8 start_led=0;
```

```
SystemInit();
RCC_Init();
JTAG_Set(1);
MouseInit();
PIDInit();
ZLG7289Init();
delay(100000);
delay(100000);
GPIO_Config1();
USART1_Config1();
NVIC_Config1();
floodwall();
GPIO_SetBits(GPIOB,GPIO_Pin_12);
while (1) {
    switch (GucMouseTask) {              /*状态机处理*/
        case WAIT:
            sensorDebug();
            delay(10000);

         if (startCheck()==true)
           {
              start++;
            }
            if(start=1&&GucGoHead)
            {
               start=0;
               GucMouseTask=START;
               delay(1000000);
            }
             break;
        case START:
        mazeSearch();
         while (1)
         {
            if (startCheck()==true)
            {
               break;
            }
         }
        break;
    default:
        break;
    }
}
}
```

# 任务二　智能鼠的坐标识别

通过前面的学习，已经知道如何控制智能鼠的运动。智能鼠在迷宫中需要自主避障寻找终点。那么智能鼠在迷宫中如何确定当前位置呢？标准竞赛迷宫共计256个单元格，如何准确地确定自身位置，成为智能鼠能否找到迷宫的必需条件。

智能鼠记录移动距离的方式有多种，比较常用的方法有两种：

## 1. 记录立柱凹槽数量

每增加一个凹槽数量，表示行走的单元格增加一个。由于立柱凹槽仅有3 mm，传感器检测数据的变化很小，所以使用这种方法对传感器的检测精度要求非常高。

## 2. 记录电动机行驶的距离

智能鼠记录电动机编码器发出的脉冲数量，经过计算得出电动机旋转的圈数，再乘以轮毂直径从而得出智能鼠行走的单元格数量。相比较而言，这种方法比较简单，准确度也较高。TQD-Micromouse-JMⅡ就是采用这种方法来获取当前位置坐标的。

## 实验一　智能鼠运行脉冲控制

核心函数1：testEncoder

```
/***********************************************************
** Function name:testEncoder
** Descriptions:编码器脉冲计数
** input parameters:无
** output parameters:无
** Returned value:无
***********************************************************/
void testEncoder(void)
{
    __GmLeft.uiPulseCtr=0;
    __GmRight.uiPulseCtr=0;
    __GucMouseState=__GOAHEAD;
    __GiMaxSpeed=SEARCHSPEED;
    __GmRight.uiPulse=5*50000;          //电动机需要运行的脉冲数
    __GmLeft.uiPulse=5*50000;
    __GmRight.cState=__MOTORRUN;
    __GmLeft.cState=__MOTORRUN;
    GuiSpeedCtr=__SPEEDUP;
    while ((__GmRight.uiPulseCtr+200 )<=__GmRight.uiPulse);
```

● 视频

实验 智能鼠运行脉冲控制

```
    while ((__GmLeft.uiPulseCtr+200)<=__GmLeft.uiPulse);
}
```

核心函数2: mouseStop

```
/***************************************************************
** Function name:mouseStop
** Descriptions:智能鼠停止程序
** input parameters:无
** output parameters:无
** Returned value:无
***************************************************************/
void mouseStop(void)
{
    __GmRight.cState=__MOTORSTOP;
    __GmLeft.cState=__MOTORSTOP;
    __GmRight.sSpeed=0;
    __rightMotorContr();
    __GmLeft.sSpeed=0;
    __leftMotorContr();
}
```

流程图。本实验通过修改电动机运行的脉冲数，来观察智能鼠运行的单元格数量，如图2-2-3所示。

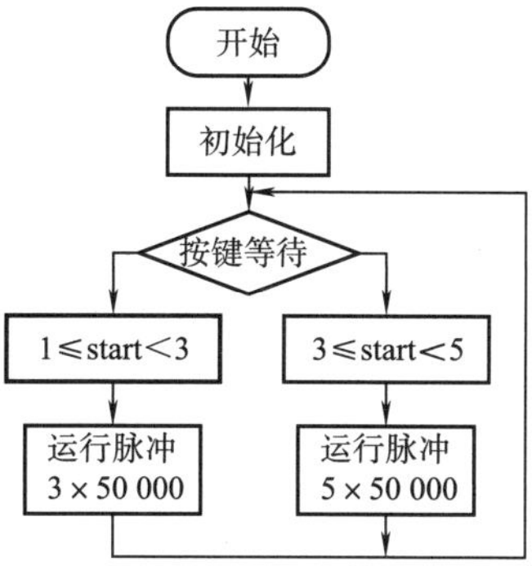

图2-2-3 智能鼠脉冲控制

主程序：

main.c

```
/***************************************************************
** Function name:main
** Descriptions:脉冲控制程序
** input parameters:无
** output parameters:无
```

```
** Returned value:无
*****************************************************************/
main (void)
{
    uint8 start=0;
    SystemInit();
    RCC_Init();
    JTAG_Set(1);
    MouseInit();
    PIDInit();
    delay(100000);
    GPIO_Config1();
    while(1) {
        switch (GucMouseTask) {              /*状态机处理*/
            case WAIT:
                sensorDebug();
                delay(10000);
                if (startCheck()==true)
                {
                    start++;
                    delay(100);
                }
                if((start<3)&&(start>=1)&&GucGoHead)
                {
                    start=0;
                    GucMouseTask=START1;
                    delay(1000000);
                }
                if((start<5)&&(start>=3)&&GucGoHead)
                {
                    start=0;
                    GucMouseTask=START2;
                    delay(1000000);
                }
                break;
            case START1:
            testEncoder1();
            mouseStop();
            while(1)
            {
                if (startCheck()==true)
                {
                    break;
                }
            }
            break;
            case START2:
```

```
            testEncoder2();
            mouseStop();
             while (1)
            {
                if (startCheck() == true)
                {
                    break;
                }
            }
            break;
        default:
            break;
        }
    }
}
```

　　尝试多次修改 __GmRight.uiPulse与__GmLeft.uiPulse的脉冲数，可以得出结论，智能鼠运行单个单元格所需的脉冲数约为51 600。

## 实验二　智能鼠坐标获取

　　当智能鼠知道自身运行的单元格数量时，再结合转弯次数与方向，就可以得出当前的坐标了。

核心函数: __mouseCoorUpdate

```
/************************************************************
** Function name:__mouseCoorUpdate
** Descriptions:根据当前方向更新坐标值
** input parameters:无
** output parameters:无
** Returned value:无
************************************************************/
void__mouseCoorUpdate (void)
{

    switch (GucMouseDir) {
    case 0:
        GmcMouse.cY++;
        break;
    case 2:
        GmcMouse.cX++;
        break;
    case 4:
        GmcMouse.cY--;
        break;
```

```
        case 6:
            GmcMouse.cX--;
            break;
        default:
            break;
    }
    __wallCheck();
    __mazeInfDebug();
}
```

流程图。本实验通过记录电动机运行的脉冲数，来确定当前智能鼠位置的坐标。流程图如图2-2-4所示。

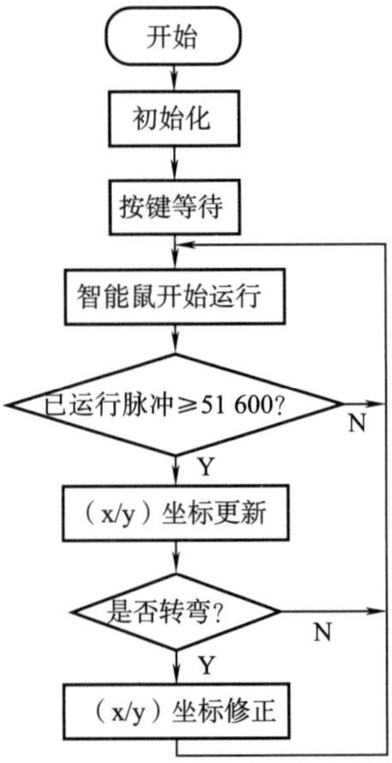

图2-2-4  更新当前位置坐标

主程序：

main.c

```
/**********************************************************
** Function name:main
** Descriptions:更新坐标主程序
** input parameters:无
** output parameters:无
** Returned value:无
**********************************************************/
main (void)
{
    uint8 n=0;                          /*有多个支路的坐标的数量*/
```

```
uint8 ucRoadStat=0;              /*统计某一坐标可前进的支路数*/
uint8 ucTemp=0;                  /*用于START状态中坐标转换*/
uint8 start=0;
uint8 start_maxspeed=0;
uint8 start_led=0;
SystemInit();
RCC_Init();
JTAG_Set(1);
MouseInit();
PIDInit();
ZLG7289Init();
delay(100000);
//__ir_Get();
delay(100000);
GPIO_Config1();
USART1_Config1();
NVIC_Config1();
floodwall();
GPIO_SetBits(GPIOB,GPIO_Pin_12);
while(1) {
    switch (GucMouseTask) {          /*状态机处理*/
        case WAIT:
            sensorDebug();
            delay(10000);

            if (startCheck()==true)
              {
                 start++;
              }
            if(start=1&&GucGoHead)
              {
                 start=0;
                 GucMouseTask=START;
                 delay(1000000);
              }
               break;
        case START:
        mazeSearch();
         while(1)
          {
             if (startCheck()==true)
              {
                 break;
              }
          }
        break;
        default:
```

```
                        break;
            }
        }
    }
```

## 任务三　智能化吸地风扇技术

智能鼠在高速运行时，会因为摩擦力不够而侧向滑动，故需要增加其对迷宫地面的抓地力。吸地风扇很好地解决了这个问题，它可以将智能鼠底板与迷宫地面之间的空气抽出，从而使内部气压减小，而外部的气压不变，从而在外部大气压下将智能鼠紧紧压在迷宫地面，使其在加减速时更加稳定。吸地风扇实物图如图2-2-5所示。

图2-2-5　吸地风扇实物图

### 一、吸地风扇设计思路及实现方法

TQD吸地风扇电动机采用专用空心杯直流电动机，扭矩大，最大转速可达50 000 r/min。控制吸地风扇驱动器采用MOSFET驱动，PWM调制，程序模块化设计，调试方便，可根据需求自主配置占空比。根据空气动力学原理，控制智能鼠底部气流压强，将智能鼠底部空气抽出，形成相对真空环境，起到增加智能鼠与地面摩擦力，克服离心力的作用，如图2-2-6所示。在吸地风扇的作用下，降低智能鼠重心，实现智能鼠高速转弯不减速，速度更快、更稳，有效减少了打滑、抖动等现象。

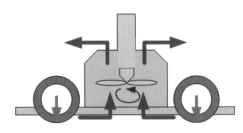

图2-2-6　吸地风扇示意图

其内外气压大小为$P_1<P_0$（$P_1$表示内压强，$P_0$表示外压强），其内外压差为$P=(P_0-P_1)=\dfrac{F}{S}$，所受摩擦力为$f=\mu N$。式中，$P$的单位为Pa；$F$的单位为N；$S$的单位为$m^2$。智能鼠摩擦力和大气压强示意图如图2-2-7所示。

图2-2-7 智能鼠摩擦力和大气压强示意图

## 二、吸地风扇电动机驱动及PWM调速

智能鼠在迷宫中运行涉及搜索、冲刺、直行、转弯，不同的运行状态需要不同的吸盘吸力来增加摩擦力。同时为了节约电池电量，需要使用PWM技术进行吸地风扇电动机调速。吸盘功率表见表2-2-1。

表 2-2-1 吸盘功率表

| PWM/% | 电压/V | 电流/mA | 功率/W | 吸起物体质量/g |
|---|---|---|---|---|
| 15 | 0.5 | 150 | 0.02 | 0 |
| 30 | 1.0 | 200 | 0.09 | 14.5 |
| 45 | 1.5 | 240 | 0.195 | 33.5 |
| 60 | 2.0 | 280 | 0.34 | 123.25 |
| 75 | 2.5 | 310 | 0.5 | 283.5 |
| 90 | 3.0 | 320 | 0.63 | 501.5 |

核心函数: fan_init

```
/***********************************************
** Function name:fan_init
** Descriptions:吸地风扇初始化
** input parameters:计数器和分频器
** output parameters:无
** Returned value:无
***********************************************/
void fan_init(u16 arr,u16 psc)          //TIM4  PB8
 {
    RCC->APB1ENR|=1<<2;         //TIM4时钟使能
  RCC->APB2ENR|=1<<3;          //PB时钟使能
  RCC->APB2ENR|=1<<0;          //AFIO时钟使能
  GPIOB->CRH&=0XFFFFFFF0;      //重置PB8
```

```
GPIOB->CRH|=0X0000000B;        //配置PB8
AFIO->MAPR&=0XFFFFEFFF;        //清除MAPR的12位
TIM4->CCR3=0;                  //初始化通道3占空比
TIM4->ARR=arr;                 //设定计数器的自动重装载值
TIM4->PSC=psc;                 //预分频器不分频
TIM4->CCMR2|=6<<4;             //CH3 PWM2模式
TIM4->CCMR2|=1<<3;             //CH3装载使能
TIM4->CCER|=1<<8;              //OC3输出使能
TIM4->CR1=0x0080;              //ARPE使能
TIM4->CR1|=0x01;               //使能定时器4
}
```

在需要调用吸盘的时候，只需要添加代码"TIM2->CCR3=数值；"即可。注意：数值不可超过800，否则会因为电流过大烧毁电路。

例如，在冲刺时开启吸盘：

```
case SPURT:
TIM2->CCR3=500;//吸盘开启
mouseSpurt();
mouseTurnback();
objectGoTo(GucXStart,GucYStart);
```

## 思考与总结

（1）如何实现智能鼠运动距离记录？

（2）如何控制智能鼠吸地风扇力度？

（3）吸地风扇技术是基于目前智能鼠速度过快、自重太轻，转弯时摩擦力不够，造成智能鼠打滑的情况而设计的。应用空气动力学原理，使用吸盘将智能鼠底部空气抽出，人为地制造上下压力差，从而增强了智能鼠的摩擦力。

# 第二篇　拓展竞技篇

前两篇中已经介绍了智能鼠的软硬件以及智能鼠的基础编程调试方法。针对IEEE国际标准智能鼠走迷宫竞赛的要求，本篇主要介绍智能鼠优化算法。掌握智能鼠走迷宫竞赛的规范，可以最快的速度完成迷宫搜索和最优路径选择；进行以往智能鼠竞赛迷宫案例要点分析，为参加IEEE国际标准智能鼠走迷宫竞赛做好准备。

项目一

智能鼠迷宫信息的获取与存储

学习目标

（1）学习智能鼠迷宫信息的获取与存储方法。

（2）学习智能鼠的路径规划和决策算法。

（3）学习智能鼠路径规划的原理。

"智能鼠"如何才能在迷宫中快速运行呢？智能鼠的主要任务是根据IEEE国际标准智能鼠走迷宫竞赛规则完成迷宫搜索和最优路径选择，是考察一个系统对一个未知环境的探测、分析及决策能力的一种比赛。下面来简单了解一下这方面的知识。

## 任务一 相对方向与绝对方向的转换

迷宫是由18 cm×18 cm大小的方格组成的，其行列各有16个方格。为了让智能鼠记住所走过的各个迷宫格的信息，就要对这256个迷宫格进行编号。很明显，用坐标是非常方便的。

规定以智能鼠放到起点时的方向为参照，此时智能鼠的正前方为Y轴正方向，后方为Y轴负方向，左方为X轴负方向，右方为X轴正方向。

为了把上下左右四个方向参数转换为微控制器能够识别的符号，在这里，将向上的方向定义为0、向右为1、向下为2、向左为3，如图3-1-1所示。

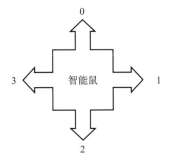

图3-1-1 迷宫的方向值定义

有了坐标和方向后，智能鼠在迷宫中行走就可以随时知道自己所处的位置和方位了，然而，对于智能鼠来说，红外线传感器的位置和方向是固定不变的，而对于迷宫来说，红外线传感器的位置和方向是随着智能鼠前进的方向变化而变化的，这就是由于参照物的不同而出现的差异。由此引出了两个方向的问题：相对方向和绝对方向。

相对方向：以智能鼠当前行走方向为参照的方向，称为相对方向。

绝对方向：以迷宫绝对坐标平面为参照的方向，称为绝对方向。

那么传感器所检测的信息如何存储才能更利于处理呢？很明显，以绝对方向存储会非常方便。这就涉及相对方向与绝对方向的转换。在这里以 Dir 记录智能鼠绝对方向值，智能鼠相对方向转换为绝对方向如表3-1-1所示。

表3-1-1　相对方向转换为绝对方向

| 相对方向 | 绝对方向 |
| --- | --- |
| 智能鼠前方 | Dir |
| 智能鼠右方 | (Dir+1)%4 |
| 智能鼠后方 | (Dir+2)%4 |
| 智能鼠左方 | (Dir+3)%4 |

有时系统还需要根据绝对方向求出相对方向，比如要控制智能鼠转向某一个绝对方向，这时就需要计算出该绝对方向处于智能鼠的哪个相对方向上，智能鼠根据相对方向来决定转向。

首先，根据目标的绝对方向（Dir_dst）和当前的绝对方向（Dir）求出方向偏差值（$\Delta$ Dir），具体公式为

$$\Delta Dir = (Dir\_dst-Dir)\%4$$

这时就可以根据方向偏差值求出智能鼠的相对方向，如表3-1-2所示。

表3-1-2　方向偏差值转换为相对方向

| 方向偏差值（$\Delta$Dir） | 相对方向 |
| --- | --- |
| 0 | 智能鼠前方 |
| 1 | 智能鼠右方 |
| 2 | 智能鼠后方 |
| 3 | 智能鼠左方 |

假设智能鼠已知当前位置坐标（$X,Y$），那么就可以求出其某绝对方向上

（相对方向可按表3-1-1转换为绝对方向）的相邻坐标值，如表3-1-3所示。该表是可逆的，即也可以根据坐标值的变化求出绝对方向。

表3-1-3　坐标转换

| 绝 对 方 向 | 相 对 方 向 |
| --- | --- |
| 当前位置 | $(X, Y)$ |
| 上方0 | $(X, Y+1)$ |
| 右方1 | $(X+1, Y)$ |
| 下方2 | $(X, Y-1)$ |
| 左方3 | $(X-1, Y)$ |

# 任务二　迷宫信息的存储方法

进行路径规划首先需要记录所有位置的挡板信息，很明显建立二维数组对整个迷宫进行坐标定义是非常有效的一个方法。每个单元格定义为一个坐标，该坐标对应的挡板信息均存储在建立的二维数组中。

当智能鼠到达一个单元格坐标时，应根据传感器检测结果记录下当前方格的挡板资料，为了方便管理和节省存储空间，每一个字节变量的低四位分别用来存储一个方格四周的挡板资料，迷宫共有16×16个方格，所以可以定义一个16×16的二维数组变量来保存整个迷宫挡板资料，如图3-1-2所示。

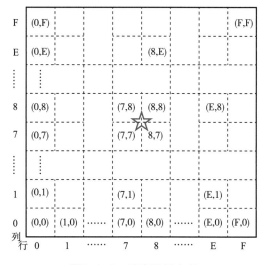

图3-1-2　迷宫坐标定义

首先将迷宫挡板资料全部初始化为0。凡是走过的迷宫格至少有一方没有挡板，也就是挡板资料不全为0，这样就可以通过单元格存储的挡板资料是否为0来确定该单元格是否曾经搜索过。挡板资料存储方式见表3-1-4。

表3-1-4　挡板资料存储方式

| 变量位 | 方位 | 是否有挡板 |
| --- | --- | --- |
| bit0 | 上方0 | 1：无挡板，0：有挡板 |
| bit1 | 右方1 | 1：无挡板，0：有挡板 |
| bit2 | 下方2 | 1：无挡板，0：有挡板 |
| bit3 | 左方3 | 1：无挡板，0：有挡板 |
| bit7～bit4 | 保留位 | |

## 思考与总结

（1）智能鼠是如何记录各单元格挡板信息的？

（2）智能鼠在运行中需要考虑当前的转弯方向和绝对方向。

### 项目二

**智能鼠路径规划的原理**

**学习目标**

（1）学习迷宫墙壁信息的存储方法。

（2）学习等高图以及转弯加权的原理，尝试在竞赛迷宫图上手绘智能鼠最短路径。

## 任务一　迷宫搜索的常用策略

迷宫搜索方法：在没有预知迷宫路径的情况下，智能鼠必须要先探索迷宫中的所有单元格，直到抵达终点为止。做这个处理的智能鼠要随时知道自己的位置及姿势，同时要记录下所有访问过的迷宫格四周是否有挡板。在搜索过程中，为了节约搜索时间，还要尽量避免重复搜索。

那么，怎样来探索迷宫呢？通常有两种策略：（1）尽快到达终点；（2）搜索整个迷宫。

这两种策略各有利弊。利用第一种策略虽然可以缩短探索迷宫所需的时间，但是不一定能够得到整个迷宫地图的资料。若找到的路不是迷宫的最优路径，这将会影响智能鼠最后冲刺的时间。利用第二种策略可以得到整个迷宫地图的资料，这样就可以求出最优路径，不过采用这种策略所使用的搜索时间较长。

## 任务二　迷宫搜索的基本法则

常用的搜索法则有三种：右手法则、左手法则和中心法则，如图3-2-1所示。

右手法则：当智能鼠的前进方向有多个可供选择时，优先向右转，其次直行，最后左转；

视　频 ●……

迷宫搜索基本
法则展示

左手法则：当智能鼠的前进方向有多个可供选择时，优先向左转，其次直行，最后右转；

中心法则：当智能鼠的前进方向有多个可供选择时，优先朝向终点的方向转动。

（a）右手法则　　　　　　　（b）左手法则　　　　　　　（c）中心法则

图3-2-1　右手法则、左手法则、中心法则

核心函数1：右手法则

```
/*************************************************************
** Function name:rightMethod
** Descriptions:右手法则，优先向右前进
** input parameters:无
** output parameters:无
** Returned value:无
*************************************************************/
void rightMethod (void)
{
    if ((GucMapBlock[GmcMouse.cX][GmcMouse.cY]&MOUSEWAY_R)&&
                                    /*智能鼠的右边有路*/
        (mazeBlockDataGet(MOUSERIGHT)==0x00)) {
                                    /*智能鼠的右边没有走过*/
        mouseTurnright();          /*智能鼠右转*/
        return;
    }
    if ((GucMapBlock[GmcMouse.cX][GmcMouse.cY]&MOUSEWAY_F)&&
                                    /*智能鼠的前方有路*/
        (mazeBlockDataGet(MOUSEFRONT)==0x00)) {
                                    /*智能鼠的前方没有走过*/
        return;                    /*智能鼠不用转弯*/
    }
    if ((GucMapBlock[GmcMouse.cX][GmcMouse.cY]&MOUSEWAY_L)&&
                                    /*智能鼠的左边有路*/
        (mazeBlockDataGet(MOUSELEFT)==0x00)) {
                                    /*智能鼠的左边没有走过*/
```

```
    mouseTurnleft();              /*智能鼠左转*/
    return;
    }
}
```

### 核心函数2：左手法则

```
/****************************************************************
** Function name:leftMethod
** Descriptions:左手法则，优先向左运动
** input parameters:无
** output parameters:无
** Returned value:无
****************************************************************/
void leftMethod (void)
{
    if ((GucMapBlock[GmcMouse.cX][GmcMouse.cY]&MOUSEWAY_L) &&
                                  /*智能鼠的左边有路*/
        (mazeBlockDataGet(MOUSELEFT)==0x00)) {
                                  /*智能鼠的左边没有走过*/
        mouseTurnleft();          /*智能鼠左转*/
        return;
    }
    if ((GucMapBlock[GmcMouse.cX][GmcMouse.cY]&MOUSEWAY_F) &&
                                  /*智能鼠的前方有路*/
        (mazeBlockDataGet(MOUSEFRONT)==0x00)) {
                                  /*智能鼠的前方没有走过*/
        return;                   /*智能鼠不用转弯*/
    }
    if ((GucMapBlock[GmcMouse.cX][GmcMouse.cY]&MOUSEWAY_R) &&
                                  /*智能鼠的右边有路*/
        (mazeBlockDataGet(MOUSERIGHT)==0x00)) {
                                  /*智能鼠的右边没有走过*/
        mouseTurnright();         /*智能鼠右转*/
        return;
    }
}
```

### 核心函数3：中心法则

```
/****************************************************************
** Function name:centralMethod
** Descriptions:中心法则，根据智能鼠目前在迷宫中所处的位置觉定使用何种搜索法则
** input parameters:无
** output parameters:无
** Returned value:无
****************************************************************/
void centralMethod (void)
{
```

```
if (GmcMouse.cX&0x08) {
    if (GmcMouse.cY&0x08) {
        /*
         *    此时智能鼠在迷宫的右上角
         */
        switch (GucMouseDir) {
        case UP:                        /*当前智能鼠向上*/
            leftMethod();               /*左手法则*/
            break;
        case RIGHT:                     /*当前智能鼠向右*/
            rightMethod();              /*右手法则*/
            break;
        case DOWN:                      /*当前智能鼠向下*/
            frontRightMethod();  /*中右法则*/
            break;
        case LEFT:                      /*当前智能鼠向左*/
            frontLeftMethod();   /*中左法则*/
            break;
        default:
            break;
        }
    } else {
        /*
         *    此时智能鼠在迷宫的右下角
         */
        switch (GucMouseDir) {
        case UP:                        /*当前智能鼠向上*/
            frontLeftMethod();   /*中左法则*/
            break;
        case RIGHT:                     /*当前智能鼠向右*/
            leftMethod();               /*左手法则*/
            break;
        case DOWN:                      /*当前智能鼠向下*/
            rightMethod();              /*右手法则*/
            break;
        case LEFT:                      /*当前智能鼠向左*/
            frontRightMethod();  /*中右法则*/
            break;
        default:
            break;
        }
    }
} else {
    if (GmcMouse.cY&0x08) {
        /*
         *    此时智能鼠在迷宫的左上角
         */
```

```
        switch (GucMouseDir) {
        case UP:                      /*当前智能鼠向上*/
            rightMethod();            /*右手法则*/
            break;
        case RIGHT:                   /*当前智能鼠向右*/
            frontRightMethod();       /*中右法则*/
            break;
        case DOWN:                    /*当前智能鼠向下*/
            frontLeftMethod();        /*中左法则*/
            break;
        case LEFT:                    /*当前智能鼠向左*/
            leftMethod();             /*左手法则*/
            break;
        default:
            break;
        }
    } else {
        /*
         *   此时智能鼠在迷宫的左下角
         */
        switch (GucMouseDir) {
        case UP:                      /*当前智能鼠向上*/
            frontRightMethod();       /*中右法则*/
            break;
        case RIGHT:                   /*当前智能鼠向右*/
            frontLeftMethod();        /*中左法则*/
            break;
        case DOWN:                    /*当前智能鼠向下*/
            leftMethod();             /*左手法则*/
            break;
        case LEFT:                    /*当前智能鼠向左*/
            rightMethod();            /*右手法则*/
            break;
        default:
            break;
        }
    }
}
```

## 任务三　等高图的制作方法

　　假设智能鼠已经搜索完整个迷宫或者只搜索了包含起点和终点的部分迷宫，且记录了已走过的每个迷宫格的挡板资料，那么它怎样根据已有信息找出一条从起点到终点最优的路径呢？下面引入等高图的概念和制作方法。

视频

等高图制作
方法

等高图在地理和气象领域有着广泛的应用。它可以标记相同高度或大小的区域范围或气压范围。在迷宫图中用等高图计算每个迷宫单元格到终点的距离，从而得到起始点到终点的距离。通过对每个单元格到终点的距离从大到小排序，就可以找出一条最优路径。

起点标记为1，根据每个单元格的挡板信息，在该单元格上标识出距离起点的最短步数，从而得到任意坐标到起点的最优路径，如图3-2-2所示。

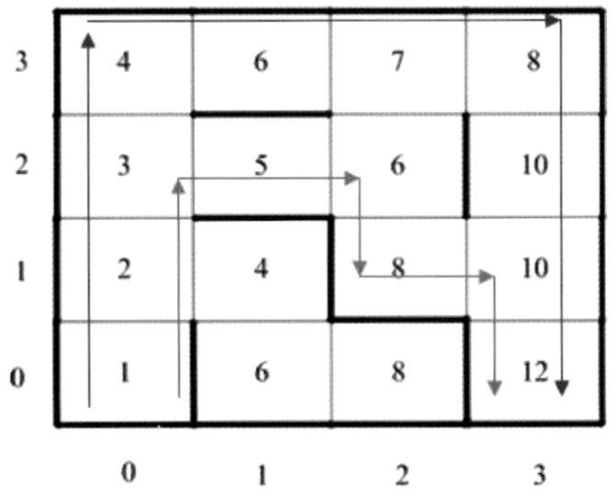

图3-2-2　等高图最终的示意图

## 思考与总结

（1）左手法则、右手法则和中心法则各有什么优缺点？

（2）智能鼠的搜索法则都是由左手法则和右手法则组合而成的。依据一定的规则，将迷宫分为若干部分，智能鼠位置不同、朝向不同，选用的搜索法则就不同。

（3）智能鼠在转弯时需要进行减速和加速，所以需要对转弯进行加权。直行优先于转弯，长直道优先于短直道。

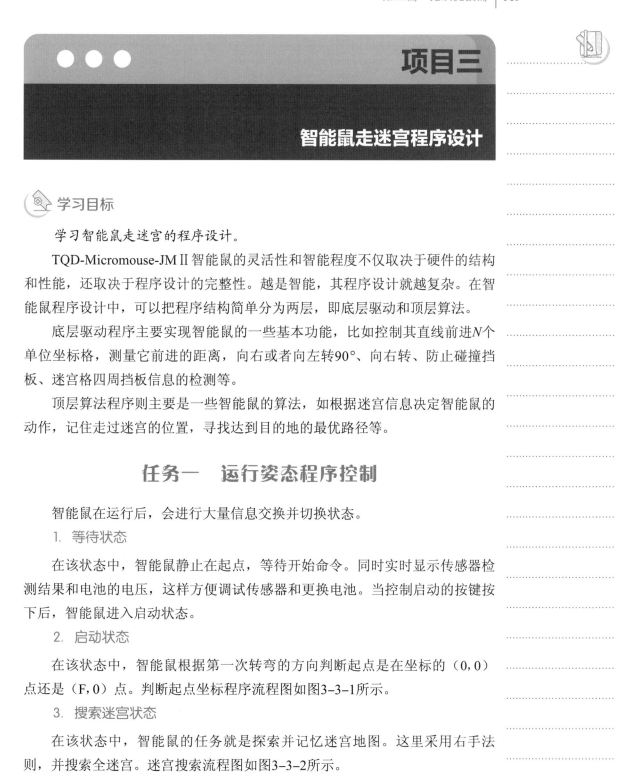

# 项目三

# 智能鼠走迷宫程序设计

🔍 **学习目标**

学习智能鼠走迷宫的程序设计。

TQD-Micromouse-JMⅡ智能鼠的灵活性和智能程度不仅取决于硬件的结构和性能，还取决于程序设计的完整性。越是智能，其程序设计就越复杂。在智能鼠程序设计中，可以把程序结构简单分为两层，即底层驱动和顶层算法。

底层驱动程序主要实现智能鼠的一些基本功能，比如控制其直线前进N个单位坐标格，测量它前进的距离，向右或者向左转90°、向右转、防止碰撞挡板、迷宫格四周挡板信息的检测等。

顶层算法程序则主要是一些智能鼠的算法，如根据迷宫信息决定智能鼠的动作，记住走过迷宫的位置，寻找达到目的地的最优路径等。

## 任务一　运行姿态程序控制

智能鼠在运行后，会进行大量信息交换并切换状态。

1. 等待状态

在该状态中，智能鼠静止在起点，等待开始命令。同时实时显示传感器检测结果和电池的电压，这样方便调试传感器和更换电池。当控制启动的按键按下后，智能鼠进入启动状态。

2. 启动状态

在该状态中，智能鼠根据第一次转弯的方向判断起点是在坐标的（0,0）点还是（F,0）点。判断起点坐标程序流程图如图3-3-1所示。

3. 搜索迷宫状态

在该状态中，智能鼠的任务就是探索并记忆迷宫地图。这里采用右手法则，并搜索全迷宫。迷宫搜索流程图如图3-3-2所示。

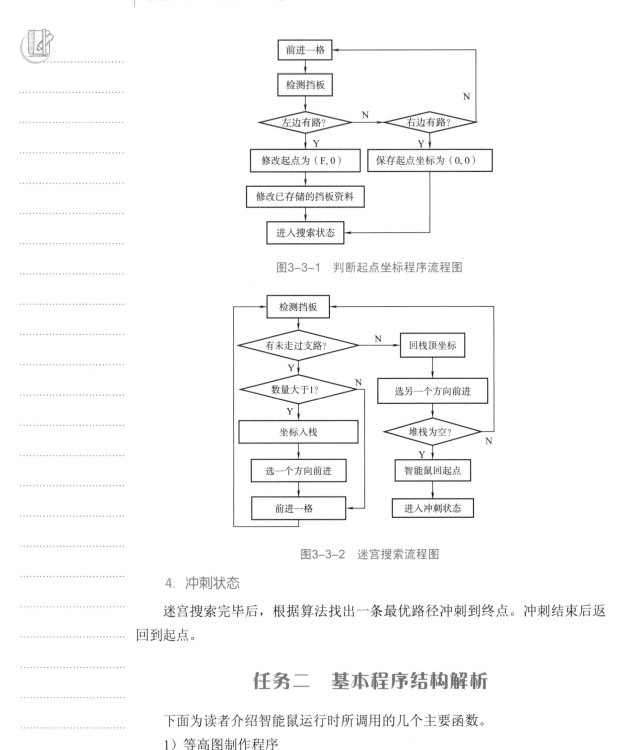

图3-3-1 判断起点坐标程序流程图

图3-3-2 迷宫搜索流程图

4. 冲刺状态

迷宫搜索完毕后，根据算法找出一条最优路径冲刺到终点。冲刺结束后返回到起点。

# 任务二 基本程序结构解析

下面为读者介绍智能鼠运行时所调用的几个主要函数。

1）等高图制作程序

智能鼠结合自身法则，整合所有坐标的挡板信息，规划最优路径。

核心函数1: mapStepEdit

```
/***************************************************************
** Function name:mapStepEdit
** Descriptions:制作以目标点为起点的等高图
** input parameters:uiX（目的地横坐标）
**                  uiY（目的地纵坐标）
** output parameters:GucMapStep[][]（各坐标上的等高值）
** Returned value:无
***************************************************************/
void mapStepEdit (int8  cX, int8  cY)
{
    uint8 n=0;                      /*有多个支路的坐标的数量*/
    uint8 ucStep=1;                 /*等高值*/
    uint8 ucStat=0;                 /*统计可前进的方向数*/
    uint8 i,j;

    GmcStack[n].cX=cX;              /*起点X值入栈*/
    GmcStack[n].cY=cY;              /*起点Y值入栈*/
    n++;
    /*
     *  初始化各坐标等高值
     */
    for (i=0; i<MAZETYPE; i++) {
        for (j=0; j<MAZETYPE; j++) {
            GucMapStep[i][j]=0xff;
        }
    }
    /*
     *  制作等高图，直到堆栈中所有数据处理完毕
     */
    while (n) {
        GucMapStep[cX][cY]=ucStep++;            /*填入等高值*/
        /*
         *  对当前坐标格里可前进的方向统计
         */
        ucStat=0;
        if ((GucMapBlock[cX][cY]&0x01)&&      /*前方有路*/
            (GucMapStep[cX][cY+1]>(ucStep))) {
                        /*前方等高值大于计划设定值，此为存储等高值*/
            ucStat++;                          /*可前进方向数加1*/
        }
        if ((GucMapBlock[cX][cY]&0x02)&&      /*右方有路*/
            (GucMapStep[cX+1][cY]>(ucStep))) {
                            /*右方等高值大于计划设定值*/
            ucStat++;                      /*可前进方向数加1*/
        }
        if ((GucMapBlock[cX][cY]&0x04)&&
```

```
                (GucMapStep[cX][cY-1]>(ucStep))) {
            ucStat++;                    /*可前进方向数加1*/
        }
        if ((GucMapBlock[cX][cY]&0x08)&&
            (GucMapStep[cX-1][cY]>(ucStep))) {
            ucStat++;                    /*可前进方向数加1*/
        }
        /*
         * 没有可前进的方向，则跳转到最近保存的分支点
         * 否则任选一可前进方向前进
         */
        if (ucStat==0) {
            n--;
            cX=GmcStack[n].cX;
            cY=GmcStack[n].cY;
            ucStep=GucMapStep[cX][cY];
        } else {
            if (ucStat > 1) {            /*有多个可前进方向，保存坐标*/
                GmcStack[n].cX=cX;    /*横坐标X值入栈*/
                GmcStack[n].cY=cY;    /*纵坐标Y值入栈*/
                n++;
            }
            /*
             * 任意选择一条可前进的方向前进
             */
            if ((GucMapBlock[cX][cY]&0x01)&&    /*上方有路*/
                (GucMapStep[cX][cY+1]>(ucStep))) {
                                        /*上方等高值大于计划设定值*/
                cY++;                    /*修改坐标*/
                continue;
            }
            if ((GucMapBlock[cX][cY]&0x02)&&
                                        /*右方有路*/
                (GucMapStep[cX+1][cY]>(ucStep))) {
                                    /*右方等高值大于计划设定值*/
                cX++;                    /*修改坐标*/
                continue;
            }
            if ((GucMapBlock[cX][cY]&0x04)&&
                                        /*下方有路*/
                (GucMapStep[cX][cY-1]>(ucStep))) {
                                    /*下方等高值大于计划设定值*/
                cY--;                    /*修改坐标*/
                continue;
            }
            if ((GucMapBlock[cX][cY]&0x08)&&    /*左方有路*/
                (GucMapStep[cX-1][cY]>(ucStep))) {
```

```
                                                    /*左方等高值大于计划设定值*/
            cX--;                                   /*修改坐标*/
            continue;
        }
    }
}
}
```

2）跳转到程序指定坐标

该程序块的作用是智能鼠按照最短路径运行到指定坐标点，当然该功能的实现前提是智能鼠已经搜索过该坐标点。

核心函数2：objectGoTo

```
/************************************************************
** Function name:objectGoTo
** Descriptions:使智能鼠运动到指定坐标
** input parameters:cXdst（目的地的横坐标）
**                  cYdst（目的地的纵坐标）
** output parameters:无
** Returned value:无
************************************************************/
void objectGoTo(int8  cXdst, int8  cYdst)
{
    uint8 ucStep=1;
    int8 cNBlock=0, cDirTemp;
    int8 cX,cY;
    GucCrossroad=0;
    cX=GmcMouse.cX;
    cY=GmcMouse.cY;
    mapStepEdit(cXdst,cYdst);              /*制作等高图*/
    /*
     *  根据等高值向目标点运动，直到达到目的地
     */
    while ((cX!=cXdst)||(cY!=cYdst)) {
        ucStep=GucMapStep[cX][cY];
        /*
         *  任选一个等高值比当前自身等高值小的方向前进
         */
        if ((GucMapBlock[cX][cY]&0x01)&&          /*上方有路*/
            (GucMapStep[cX][cY+1]<ucStep)) {      /*上方等高值较小*/
        cDirTemp=UP;                       /*记录方向*/
        if (cDirTemp==GucMouseDir) {
                                    /*优先选择不需要转弯的方向*/
            cNBlock++;                     /*前进一个方格*/
            cY++;
```

```
                                if((GucMapBlock[cX][cY]&0x0f)==0x0f)
                                  GucCrossroad++;
                                continue;               /*跳过本次循环*/
                            }
                        }
                        if ((GucMapBlock[cX][cY]&0x02) &&        /*右方有路*/
                            (GucMapStep[cX+1][cY]<ucStep)) {
                                                                 /*右方等高值较小*/
                            cDirTemp=RIGHT;                 /*记录方向*/
                            if (cDirTemp==GucMouseDir) {
                                                    /*优先选择不需要转弯的方向*/
                                cNBlock++;                  /*前进一个方格*/
                                cX++;
                                if((GucMapBlock[cX][cY]&0x0f)==0x0f)
                                  GucCrossroad++;
                                continue;               /*跳过本次循环*/
                            }
                        }
                        if ((GucMapBlock[cX][cY]&0x04) &&    /*下方有路*/
                            (GucMapStep[cX][cY-1]<ucStep)) {     /*下方等高值较小*/
                            cDirTemp=DOWN;                  /*记录方向*/
                            if (cDirTemp==GucMouseDir) {  /*优先选择不需要转弯的方向*/
                                cNBlock++;                  /*前进一个方格*/
                                cY--;
                                if((GucMapBlock[cX][cY]&0x0f)==0x0f)
                                  GucCrossroad++;
                                continue;               /*跳过本次循环*/
                            }
                        }
                        if ((GucMapBlock[cX][cY]&0x08) &&        /*左方有路*/
                            (GucMapStep[cX-1][cY]<ucStep)) {     /*左方等高值较小*/
                            cDirTemp=LEFT;                  /*记录方向*/
                            if (cDirTemp==GucMouseDir) {  /*优先选择不需要转弯的方向*/
                                cNBlock++;              /*前进一个方格*/
                                cX--;
                                if((GucMapBlock[cX][cY] & 0x0f)==0x0f)
                                GucCrossroad++;
                                continue;               /*跳过本次循环*/
                            }
                        }
                        cDirTemp=(cDirTemp+8-GucMouseDir)%8;    /*计算方向偏移量*/
                        GucDirTemp=cDirTemp;
                        if (cNBlock) {
                          if((GucCrossroad<=1)&&(cNBlock>1))
                              mouseGoahead_L(cNBlock);        /*前进cNBlock步*/
                          else{
                            mouseGoahead_L(cNBlock);
```

```
            GucCrossroad=0;
        }


        }
    cNBlock = 0;
    /*任务清零*/
    /*
     *控制智能鼠转弯
     */
    switch (cDirTemp) {
    case 2:
        mouseTurnright_C();
        break;
    case 4:
        mouseTurnback();
        break;
    case 6:
        mouseTurnleft_C();
        break;
    default:
        break;
    }
    GmcMouse.cX=cX;
    GmcMouse.cY=cY;
}
/*
 *   判断任务是否完成, 否则继续前进
 */
    if (cNBlock) {
        if((GucCrossroad<=1)&&(cNBlock>1))
          mouseGoahead_L(cNBlock);              /*前进cNBlock步*/
        else{
          mouseGoahead_L(cNBlock);
          GucCrossroad=0;
        }
        GmcMouse.cX=cX;
        GmcMouse.cY=cY;
    }
}
```

3) 统计未搜索的支路

该程序用于统计指定坐标四周存在的还未搜寻过的支路总数, 以供系统运用搜索策略。程序如下:

核心函数3: crosswayCheck

```
/******************************************************************
** Function name:crosswayCheck
** Descriptions:统计某坐标存在还未走过的支路数
** input parameters:ucX（需要检测点的横坐标）
**                  ucY（需要检测点的纵坐标）
** output parameters:无
** Returned value:ucCt（未走过的支路数）
******************************************************************/
uchar crosswayCheck (char  cX, char  cY)
{
    uchar ucCt=0;
    if ((GucMapBlock[cX][cY]&0x01)&&
                                        /*绝对方向，迷宫上方有路*/
        (GucMapBlock[cX][cY+1])==0x00) {
                                           /*绝对方向，迷宫上方未走过*/
        ucCt++;
                                              /*可前进方向数加1*/
    }
    if ((GucMapBlock[cX][cY]&0x02)&&
                                        /*绝对方向，迷宫右方有路*/
        (GucMapBlock[cX+1][cY])==0x00) {
                                           /*绝对方向，迷宫右方没有走过*/
        ucCt++;                            /*可前进方向数加1*/
    }
    if ((GucMapBlock[cX][cY]&0x04)&&
                                        /*绝对方向，迷宫下方有路*/
        (GucMapBlock[cX][cY-1])==0x00) {
                                           /*绝对方向，迷宫下方未走过*/
        ucCt++;

                                              /*可前进方向数加1*/
    }
    if ((GucMapBlock[cX][cY]&0x08)&&
                                        /*绝对方向，迷宫左方有路*/
        (GucMapBlock[cX-1][cY])==0x00) {
                                           /*绝对方向，迷宫左方未走过*/
        ucCt++;
                                              /*可前进方向数加1*/
    }
    return ucCt;
}
```

## 4）TQD-Micomouse-JM Ⅱ 走迷宫主程序

```
/******************************************************************
** Function name:main
```

```
** Descriptions:主函数
** input parameters:无
** output parameters:无
** Returned value:无
********************************************************/
main (void)
{
    uint8 n=0;                    /*有多个支路的坐标的数量*/
    uint8 ucRoadStat=0;           /*统计某一坐标可前进的支路数*/
    uint8 ucTemp=0;               /*用于START状态中坐标转换*/
    uint8 start=0;
    uint8 start_maxspeed=0;
    uint8 start_led=0;
    SystemInit();
    RCC_Init();
    JTAG_Set(1);
    MouseInit();
    PIDInit();
    ZLG7289Init();
    delay(100000);
    delay(100000);
    GPIO_Config1();
    USART1_Config1();
    NVIC_Config1();
    floodwall();
    GPIO_SetBits(GPIOB,GPIO_Pin_12);
    while (1) {
        switch (GucMouseTask) {       /*状态机处理*/
            case WAIT:
                sensorDebug();
                delay(10000);

             if (startCheck()==true)
               {
                   start++;
                }
               if((start==1)&&GucGoHead)//&&GucGoHead
               {
                   start=0;
                   zijiaozheng0=1;

                    GPIO_ResetBits(GPIOB,GPIO_Pin_12);
            delay(50000000);
            GucMouseTask=START;
                delay(1000000);
                GPIO_SetBits(GPIOB,GPIO_Pin_12);
                zijiaozheng0=1;
```

```
                            }
                            if((start<5)&&(start>=3)&&GucGoHead)
                            {

                                GPIO_ResetBits(GPIOB,GPIO_Pin_12);
                                zijiaozheng0=1;
                                 start=0;
                                wallget();
                                delay(10000000);
                                StartGet();
                                delay(10000000);
                                delay(1000000);
                                delay(1000000);delay(1000000); delay(1000000);
                                GucMouseTask=SPURTL;
                                 delay(1000000);
                            }
                            break;
                    case START:                  /*判断智能鼠起点的横坐标*/
                        GPIO_ResetBits(GPIOB,GPIO_Pin_12);
                      mazeSearch();           /*向前搜索*/
                          if (GucMapBlock[GmcMouse.cX][GmcMouse.cY] & 0x08)
                          {
                              if (MAZETYPE==16)
                              {
                                  GucXGoal0=8;
                                  GucXGoal1=7;
                              }
                              GucXStart=MAZETYPE-1;
                              GmcMouse.cX=MAZETYPE-1;
                              ucTemp=GmcMouse.cY;
                              do {
                          GucMapBlock[MAZETYPE-1][ucTemp]=GucMapBlock[0][ucTemp];
                          GucMapBlock0[MAZETYPE-1][ucTemp]=GucMapBlock0[0][ucTemp];
                                  GucMapBlock0[0][ucTemp]=0;
                          GucMapBlock1[MAZETYPE-1][ucTemp]=GucMapBlock1[0][ucTemp];
                                  if(ucTemp>0)
                                  {
                          GucMapBlock1[MAZETYPE-2][ucTemp-1]=0x1d;
                                  }
                                  GucMapBlock1[0][ucTemp+1]=0x17;
                                  GucMapBlock1[1][ucTemp]=0x1f;
                                  GucMapBlock[0][ucTemp]=0x10;
                                  GucMapBlock[1][ucTemp]=0x10;
                              }while (ucTemp--);
                              GucMapBlock1[0][0]=0x13;
                              GucMapBlock1[1][0]=0x1b;
                              GucMapBlock1[MAZETYPE-2][0]=0x19;
```

```
                // 在OFFSHOOT[0]中保存起点坐标
                GmcCrossway[n].cX=MAZETYPE-1;
                GmcCrossway[n].cY=0;
                n++;
                GucMouseTask=MAZESEARCH;
            }
            if (GucMapBlock[GmcMouse.cX][GmcMouse.cY]&0x02)
            {
                // 在OFFSHOOT[0]中保存起点坐标
                GmcCrossway[n].cX=0;
                GmcCrossway[n].cY=0;
                n++;
                GucMouseTask=MAZESEARCH;
            }
        break;
    case MAZESEARCH:
        centralMethodnew();
        goalwall();
        StartSave();
        wallsave();
        mouseTurnback();
        objectGoTo1(GucXStart,GucYStart);
        onestep3();
        mouseTurnbackqi();
        GucMouseTask=SPURTL;
        break;
    case  SPURTL:
    TIM4->CCR3=600;delay(1000000); delay(1000000);delay(1000000);
//设置占空比，延时
        StartGet();
        wallget();
        goalwall();
        mouseSpurt_CC();
        onestep();
        mouseTurnback();
        objectGoTo1(GucXStart,GucYStart);
        onestep();
        mouseTurnbackqidian();
        GucMouseTask=SPURT45;
        break;
    case SPURT45:
        TIM4->CCR3=600;
        mouseSpurt_45();
        usepiancha=1;
        onestep1();
        mouseTurnback();
        __GmSPID.sRef=145;
```

```
                    TIM4->CCR3=0;
                    objectGoTo1(GucXStart,GucYStart);
                    mouseTurnbackqi();
                    while (1)
                    {
                        if (startCheck()==true)
                        {
                            break;
                        }
                    }
                    break;
            default:
                    break;
            }
        }
    }
```

## 思考与总结

（1）智能鼠程序设计各个阶段有何联系？

（2）智能鼠主要函数分别起什么作用？

（3）智能鼠走迷宫是由多个主要函数配合执行的，只有都调试准确，才能使智能鼠顺利通过迷宫到达终点。

附　　录

TODD MICROMOUSE I/O / JD / JM

# 附录A

## 风靡全球的国际智能鼠走迷宫竞赛

2019年是智能鼠走迷宫竞赛有史以来，最具兴盛发展、硕果累累的一年，在世界各地如火如荼地举行国际智能鼠走迷宫竞赛如图A–1所示。

1月，在印度孟买举办印度智能鼠国际竞赛。

3月，在美国加利福尼亚州举办APEC国际智能鼠竞赛。

4月，在葡萄牙Gondomar（贡多马尔）举办国际智能鼠走迷宫竞赛。

5月，在中国天津举办IEEE智能鼠走迷宫国际邀请赛。

6月，在英国伦敦举办智能鼠国际竞赛。

8月，在智利举办智能鼠走迷宫国际竞赛。

10月，在埃及举办埃及智能鼠国际竞赛。

11月，在日本东京举办全日本智能鼠国际公开赛。

图A–1　国际智能鼠竞赛赛事安排

国际智能鼠走迷宫竞赛将成为全球高等教育、职业教育、普通教育，技术创新产教融合发展的助推器。在人工智能智能鼠走迷宫竞赛蓬勃发展的国际大环境下，教育领域适时地引进国际知名赛事提升学生的专业综合能力，掌握实践与创新的经验，助力产教融合发展，为行业、产业、企业培养更多优秀种子人才。

1. 中国IEEE智能鼠走迷宫国际邀请赛

从2009年开始，天津启诚伟业科技有限公司把智能鼠走迷宫竞赛引入中国，将IEEE智能鼠走迷宫竞赛进行本土化创新改革，对满足产业优化升级，开阔国际视野，掌握实践与创新经验，培育高技术高技能人才，起到了引领推动作用。

从2016年至2019年，连续四届举办中国IEEE智能鼠走迷宫国际邀请赛，该竞赛由天津市教育委员会主办，天津启诚伟业科技有限公司和天津渤海职业技术学院承办，如图A-2所示。

图A-2    从2016年开始举办"中国IEEE智能鼠走迷宫国际邀请赛"

● 视频

中国IEEE智能鼠走迷宫国际邀请赛

目前，中国IEEE智能鼠走迷宫国际邀请赛设置了"中学、高职、本科、硕士、职业"共五个竞赛组别，旨在提升大赛的社会参与度和专业覆盖面。智能鼠走迷宫竞赛已经发展成为了系统化培养和教育的重要载体。充分体现光机电结合、软硬件结合、控制与机械结合，演绎"工程"课程概念的同时，延伸和扩展了"创新"课程的理念，使得学生的学习内容和教师的授课方式都有了全新的内涵，真正着眼于综合素质的培养，创造快乐素质教育。

中国IEEE智能鼠走迷宫国际邀请赛主要特点：

（1）参赛群体：既面对在校大学生，也面对小学、中学和职业人士，体现贯通式培养，终身教育的特点。还包括国际智能鼠专业级选手和历届国际智能鼠竞赛获奖选手。

（2）迷宫场地：既有面向中小学的8×8智能鼠迷宫场地，也有面向大专院校的16×16全迷宫古典智能鼠场地；更有面对精英选手的25×32半尺寸智能鼠迷宫场地。体现竞赛的延展性，以智能鼠走迷宫竞赛为核心形式，不同学习阶段的学生都可以参赛。

（3）竞赛项目：既有智能鼠走迷宫赛项，又有自走车赛项，体现了竞赛既有技术性也有工程性，以工程应用为导向的竞赛思想。

（4）竞赛规则：普通教育、职业教育、高等教育、职业精英竞赛规则的相同点和差异点对比见表A-1。

表A-1　竞赛规则相同点和差异点对比

| 参赛类别 | 普通教育组 | 职业教育组 | 高等教育组 | 职业精英组 |
|---|---|---|---|---|
| 竞赛形式 | （1）程序参数App在线调试。<br>（2）图形化趣味编程。<br>（3）IOT智能传感技术应用。<br>（4）8×8迷宫竞赛 | （1）理论知识考核。<br>（2）根据裁判现场任务编程并实现相应功能。<br>（3）现场技术答辩。<br>（4）16×16古典迷宫竞速 | （1）DIY外观及结构机械设计。<br>（2）硬件技术创新。<br>（3）程序算法创新。<br>（4）16×16古典迷宫竞速 | （1）DIY外观及结构机械设计。<br>（2）硬件技术创新。<br>（3）程序算法创新。<br>（4）25×32半尺寸迷宫竞速 |
| 竞赛内容 | （1）组装任务10%。<br>（2）调试任务40%。<br>（3）竞速任务50% | （1）理论考核20%。<br>（2）创新赛30%。<br>（3）竞速赛50% | （1）创新赛20%。<br>（2）竞速赛80% | 竞速赛100% |

智能鼠国际专家现场培训指导如图A-3所示。

图A-3　智能鼠国际专家现场培训指导

2. 美国APEC世界智能鼠竞赛

1977年在美国纽约举行的首场令人震撼的智能鼠走迷宫竞赛，由IEEE与APEC共同主办。于是诞生了国际上最有影响力的美国APEC世界智能鼠竞赛。号称智能鼠世界三大赛事之一，截止到2019年已经举办了34届。

APEC组织的官网网址：http://www.apec-conf.org/。

美国智能鼠爱好者的网址：http://micromouseusa.com/，如图A-4所示。

（a）　　　　　　　　　　　　（b）

图A-4　美国APCE世界智能鼠竞赛官网截屏

竞赛时间：每年2月到4月之间。

竞赛地点：每年不同（举办过的地点包括北卡罗来纳州、德克萨斯州、佛罗里达州、加利福尼亚州等），每年都会有来自美国、英国、日本、韩国、新加坡、印度、中国等国家的选手踊跃参赛，如图A-5所示。

图A-5　中国选手参加第30届美国APEC世界智能鼠竞赛

### 3. 英国智能鼠国际竞赛

从1980年至今，英国智能鼠国际竞赛已经成长为国际知名的智能鼠竞赛之一。

竞赛时间：每年6月。

竞赛地点：英国伯明翰城市大学。

该项竞赛由英国智能鼠和机器人协会主办，英国的智能鼠竞赛特点在于重在参与，从中学生、大学生、到社会人员，任何人都可以参赛。所有的参赛队员分为不同的组别，迷宫难度也适当调整。该项竞赛分为line follower、wall follower、maze solver等项目，吸引了来自全球10余个国家，50余支队伍参赛。

英国智能鼠国际竞赛评分规则介绍：在16×16的迷宫中，参赛智能鼠需要完成起点到终点的搜索和全迷宫的遍历，求解最佳路线并完成由起点到终点的冲刺。计分时间=搜索时间（第一次搜索到终点的时间）/30+冲刺时间（完成起点到终点最短路径的高速冲刺）+惩罚时间（撞挡板罚时：3 s/次）。

官方网址：https://ukmars.org/index.php/Main_Page，如图A-6所示。

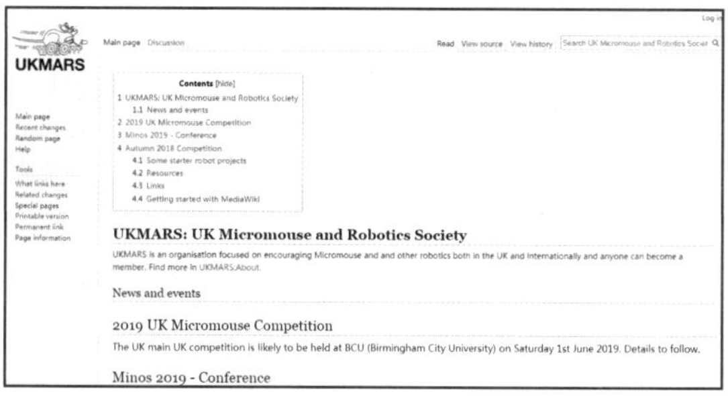

图A-6　英国智能鼠国际竞赛官网截屏

### 4. 全日本智能鼠国际公开赛

全日本智能鼠国际公开赛从1980年到2019年已经举办40届。

竞赛时间：每年的11月底或12月初。

竞赛地点：日本东京。

官方网址：http://www.ntf.or.jp/mouse/Micromouse2018/index.html，如图A-7所示。

图A-7　全日本智能鼠国际公开赛官网截屏

视频

全日本
Micromouse
国际公开赛

每年竞赛都有来自美国、英国、日本、新加坡、中国、蒙古、智利、葡萄牙等二十多个国家的智能鼠参赛队角逐（见图A-8）。

赛项由古典智能鼠赛项、半尺寸智能鼠赛项、自走车赛项组成。参赛队由中学生、大学生和职业精英组成，据统计有300多支参赛队。全日本智能鼠国际公开赛可以说是代表当今，国际智能鼠技术领域级别最高、技术最强的赛事，所以备受瞩目。

图A-8　第39届全日本智能鼠国际公开赛颁奖照片

### 5. 智利智能鼠走迷宫国际竞赛

智利外交部希望通过国际智能鼠走迷宫竞赛，推动智利青少年科技创新以及国际间技术创新和交流合作，从而带动智利经济发展。2018年12月3日在全日本智能鼠国际公开赛期间，智利驻日本大使馆主持召开"智利智能鼠走迷宫国际竞赛研讨会"特别邀请国际智能鼠专家（美国的David Otten、英国的Peter Harrison、日本的中川友纪子、中国的宋立红、智利的Benjamin等）共同商议智利智能鼠走迷宫国际竞赛统一标准和规范，如图A-9、图A-10所示。

图A-9　智利外交部会议——共同探讨智能鼠发展

图A-10　2018年智利智能鼠走迷宫国际竞赛规则研讨会

### 6. 葡萄牙智能鼠走迷宫国际竞赛

2019年4月27日在Gondomar（贡多马尔）举办，由葡萄牙杜罗大学技术执行委员会主办。

葡萄牙竞赛开始于2011年，旨在通过培养创造力和能力来提供完整的技术学习环境，迄今已经成功举办9届。

竞赛时间：每年4月或5月。

竞赛地点：葡萄牙。

官方网址：http://www.micromouse.utad.pt/，如图A-11所示。

2019年4月27日，当地时间18时整，在葡萄牙波尔图体育馆中，来自英国、中国、葡萄牙、西班牙、巴西、新加坡等国家的参赛队，正在上演一场紧张激烈的国际智能鼠走迷宫竞赛。伴随着中国智能鼠稳健搜索和极速冲刺，掌声欢呼声在葡萄牙波尔图体育馆雷鸣般响起……启诚智能鼠实现突破性成果，取得了世界亚军的殊荣（见图A-12）。

视 频

葡萄牙
Micromouse
走迷宫国际
竞赛

图A-11　葡萄牙智能鼠走迷宫
国际竞赛官网截屏

图A-12　启诚智能鼠荣获葡萄牙
智能鼠走迷宫竞赛世界亚军

赛后葡萄牙智能鼠走迷宫竞赛组委会主席安东尼奥表示，近年来中国的综合国力和技术实力不断增强，特别是教育领域对于科技创新和工程素养越来越重视。启诚智能鼠首次参加葡萄牙智能鼠走迷宫竞赛，就获得了优异成绩非常可喜可贺（见图A-13）。

图A-13　中国及葡萄牙智能鼠走迷宫竞赛专家现场技术交流

### 7. 印度智能鼠国际竞赛

2020年1月4日，印度孟买举办"2020年第23届亚洲科技节首届智能鼠国际大赛"，来自印度、中国、澳大利亚、尼泊尔、斯里兰卡、孟加拉等国家的代表队参加本届大赛（见图A-14）。中国天津智能鼠代表队力克群雄，以绝对优势包揽"金、银、铜"全部奖牌，将17.5万卢比奖金完美收入囊中。

竞赛时间：每年1月份。

竞赛地点：印度孟买。

官方网址：http://techfest.org/competitions/Micromouse。

图A-14　印度智能鼠国际竞赛合影

特别值得一提的是，印度金奈理工学院鲁班工坊代表队（见图A-15）采用2017年中方赠送的IEEE国际标准智能鼠走迷宫创新型教学设备TQD-Micromouse-JD智能鼠参加本届大赛，荣获印度国内大赛冠军、世界精英组第四名的好成绩，并赢得5 000卢比奖金，成为印度国际智能鼠走迷宫竞赛的明星赛队。印度鲁班工坊指导教师卡西克表示，金奈理工学院鲁班工坊代表队能取得这样优异的成绩，是三年来鲁班工坊的师生和支持企业（启诚科技）共同努力的成果。

图A-15　印度鲁班工坊参赛师生合影

### 8. 埃及智能鼠国际竞赛

埃及国际电气电子工程师学会（IEEE）现在已经发展成为具有较大影响力的国际学术和技术组织之一。30多年来，一直在推动和指导电气电子技术的发展与创新。这项技术包括电子元件、电路理论和设计技术的应用，以及针对有效转换、控制和电力状况分析工具的开发。IEEE成员包括杰出的研究人员、从业人员和杰出的获奖者。

图A-16所示为埃及IEEE在官网首页为智能鼠竞赛做的宣传。IEEE电力电子和可再生能源大会为颇具亮点的国际智能鼠大赛优胜者准备了丰厚的奖金。特等奖相当于1 000美元；杰出表现奖相当于700美元；最佳创新设计奖相当于500美元。参赛队来自埃及国内或国际工程学或相关专业的学生，也可以是高中生。每个参赛队中最多允许有两名学生。

官方网址：http://www.ieee-cpere.org/International_Competition.html。

图A-16　埃及智能鼠国际竞赛

埃及智能鼠国际竞赛纪实如图A-17所示。

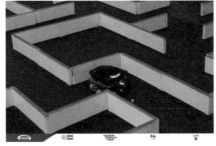

图A-17　埃及智能鼠国际竞赛纪实照片

## 附录B

### 高级段经典竞赛案例分析

IEEE国际标准智能鼠走迷宫竞赛具有一定难度，是一项富有挑战性和趣味性的学生比赛，在国内外享有一定的知名度和影响力。智能鼠走迷宫竞赛项目，从技术上涵盖了物联网应用技术、电子信息工程技术、嵌入式技术、通信技术、软件技术、计算机网络技术、信息安全技术、移动通信技术、计算机应用技术、应用电子技术、计算机控制技术、机电一体化技术、自动化技术等多个专业技术，涉及传感器检测、人工智能、自动控制和机电运动部件应用等技能和综合职业素养。全面展现高等教育和职业教育的发展水平，可提高电子信息类高素质、高技能应用型人才的培养质量。

通过竞赛，推动了电子信息行业企业人才能力需求，顺应科技发展将嵌入式技术开发、智能算法优化等前端先进技术融入竞赛内容中，进一步深化校企合作，引导电子信息类专业开展单片机应用、嵌入式技术应用、物联网技术应用等。课程建设和教学改革，促进创新型人才培养模式的发展，增强了电子信息类专业学生就业竞争力，推进创新创业教育，强化创业指导和服务，提高就业水平。

随着时代的发展，科技的进步，智能鼠顺应现代科技发展，经过多年的蜕变与优化，已经成为集人工智能、嵌入式、智能传感等融合新技术于一体的优秀实训教育平台。

四十多年来，IEEE每年举办一次国际性的智能鼠走迷宫竞赛。自举办以来，各个国家和地区的学生踊跃参加，尤其是美国和欧洲国家的高校学生，为此有些大学还特别开设了"智能鼠原理与制作"的选修课程，如图B-1、图B-2所示。

下面就以比较有代表性的国内外智能鼠走迷宫竞赛典型迷宫地图为例进行解析说明。

#### 1. 美国APEC世界Micromouse竞赛迷宫分析

2015年3月16日，第30届美国APEC世界Micromouse竞赛成功举办。天津启诚伟业科技有限公司带领天津大学生联合代表队参加本次竞赛，让天津智能鼠技术与国际接轨，是天津智能鼠历史性的转折，具有里程碑式的意义。

图B-1　组委会主席David Otten教授

图B-2　日本选手宇都宫正和

　　本次届竞赛是一届非常成功的赛事。传感器从数字型向模拟型过渡，运动结构也从步进电动机向空心杯直流电动机发展，在这次赛事上出现了融合吸地风扇的智能鼠。

　　本次竞赛所采用的迷宫整体难度上相对均衡，有众多的路径可供选择，既有长直道展现智能鼠的高速运动性能，也有适中的连续转弯体现智能鼠的精确控制，如图B-3所示。

 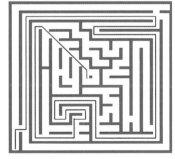

（a）迷宫原图　　　　　　　　（b）最佳路径

图B-3　迷宫图解析

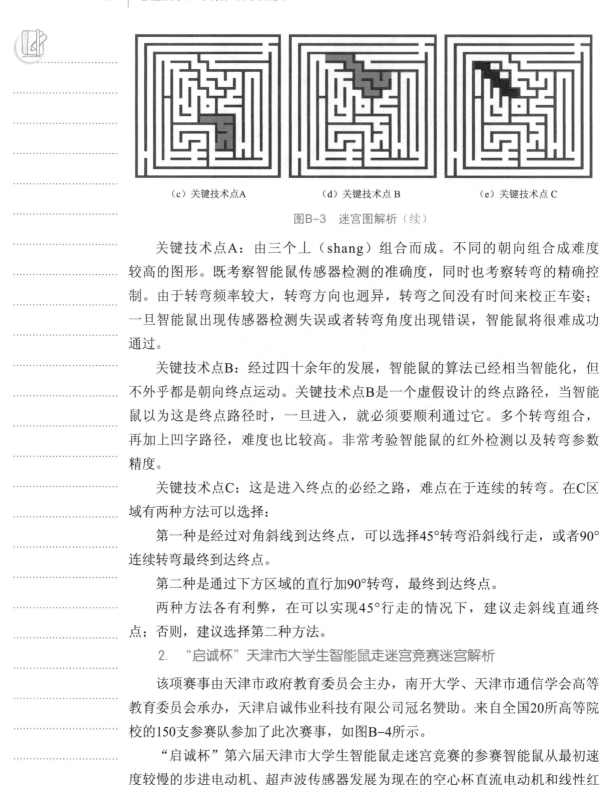

（c）关键技术点A          （d）关键技术点B          （e）关键技术点C

图B-3　迷宫图解析（续）

关键技术点A：由三个⊥（shang）组合而成。不同的朝向组合成难度较高的图形。既考察智能鼠传感器检测的准确度，同时也考察转弯的精确控制。由于转弯频率较大，转弯方向也迥异，转弯之间没有时间来校正车姿；一旦智能鼠出现传感器检测失误或者转弯角度出现错误，智能鼠将很难成功通过。

关键技术点B：经过四十余年的发展，智能鼠的算法已经相当智能化，但不外乎都是朝向终点运动。关键技术点B是一个虚假设计的终点路径，当智能鼠以为这是终点路径时，一旦进入，就必须要顺利通过它。多个转弯组合，再加上凹字路径，难度也比较高。非常考验智能鼠的红外检测以及转弯参数精度。

关键技术点C：这是进入终点的必经之路，难点在于连续的转弯。在C区域有两种方法可以选择：

第一种是经过对角斜线到达终点，可以选择45°转弯沿斜线行走，或者90°连续转弯最终到达终点。

第二种是通过下方区域的直行加90°转弯，最终到达终点。

两种方法各有利弊，在可以实现45°行走的情况下，建议走斜线直通终点；否则，建议选择第二种方法。

2. "启诚杯"天津市大学生智能鼠走迷宫竞赛迷宫解析

该项赛事由天津市政府教育委员会主办，南开大学、天津市通信学会高等教育委员会承办，天津启诚伟业科技有限公司冠名赞助。来自全国20所高等院校的150支参赛队参加了此次赛事，如图B-4所示。

"启诚杯"第六届天津市大学生智能鼠走迷宫竞赛的参赛智能鼠从最初速度较慢的步进电动机、超声波传感器发展为现在的空心杯直流电动机和线性红

附   录 | 115

外传感器。参赛选手的水平也在逐年提高。在2017年举办的竞赛中，采用的是一个非常经典的迷宫，难度高、开放性强是它最主要的特点，如图B-5所示。

图B-4　2017年"启诚杯"智能鼠竞赛开幕式

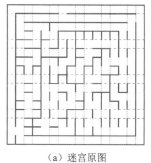

（a）迷宫原图　　　　　　　　　　（b）最佳路径

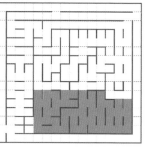

（c）关键技术点A　　　　（d）关键技术点 B　　　　（e）关键技术点 C

图B-5　迷宫图解析

　　关键技术点A：由于第一个路口紧贴起点位置，使用智能算法的智能鼠通常都不会在这里右转，这也就造成了智能鼠沿着最外侧挡板行走一整圈后又回到了这一路口，最终进入A区域。几乎完全对称的图案，众多的路口，使得智能鼠几乎没有机会进行车姿校正。这对智能鼠的性能是一个巨大的考验。

　　关键技术点B：与关键技术点A相同，B区域也是以高开放性为主。众多的路口，使得智能鼠没有足够的时间进行车姿校准，同时又给了高性能智能鼠45°走斜线的机会。智能鼠的性能差距，很容易判断出参赛选手水平高低。

关键技术点C：C区域是进入终点的必经之路，使用智能算法的智能鼠进入这一区域后，通常不会去行走"死胡同"，直接进入终点完成竞赛。这里也成为区分算法是否智能的区域。

3. 第二届IEEE智能鼠走迷宫国际邀请赛迷宫解析

2017年8月，第二届"中国IEEE智能鼠走迷宫国际邀请赛"成功举办。本届大赛吸引了来自英国、泰国、蒙古等国家代表队，以及中国实力雄厚的天津大学、南开大学、北京交通大学、天津中德应用技术大学等智能鼠精英赛队参加，如图B-6所示。

图B-6　第二届IEEE智能鼠走迷宫国际邀请赛

本次竞赛采用了一幅比较有特色的迷宫。随着智能算法的流行，大型竞赛所选用的迷宫，越来越注重为智能算法增加难度系数，如图B-7所示。

关键技术点A：蓝色箭头所在位置是智能算法的首选路口。这一部分是一个完全封闭的区域，难度也较低；但是当智能鼠进入这一区域，就需要面临大量的无用搜索和转弯。只有成功搜索完毕并顺利退出A区域，智能鼠才能进入其他区域的搜索。

关键技术点B：B区域包含智能鼠到达终点的两条路径。大量的路口以及左转右转结合，非常注重考察智能鼠红外检测精度和转弯角度的准确性。智能鼠在任何一个路口发生传感器误判或者转弯精度有误都将是致命的。

关键技术点C：C区域同样是到达终点的一条路径。阶梯状的连续转弯，非常考验智能鼠的转弯准确度。在可以实现45°斜直线行走的情况下，推荐选择这一路径。

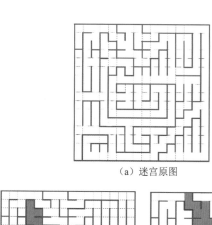 

(a) 迷宫原图 　　　　　　　(b) 最佳路径

 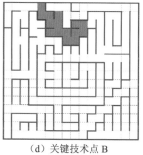 

(c) 关键技术点A　　　　　(d) 关键技术点 B　　　　　(e) 关键技术点 C

图B-7　迷宫解析

## 4. 世界智能鼠经典赛事迷宫范例（见图B-8~图B-10）

图B-8　2012年全日本智能鼠国际公开赛（新生赛和专家赛）

 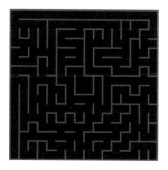

图B-9　2000年英国UK智能鼠 　　　　图B-10　2002年美国APEC
　　　　国际邀请赛 　　　　　　　　　　　　智能鼠国际大赛

# 附录C

## TQD-Micromouse-JMⅡ器件清单

TQD-Micromouse-JMⅡ器件清单见表C-1。

表C-1 TQD-Micromouse-JMⅡ器件清单

| 序　号 | 名　称 | 数　量 | 备　注 |
|---|---|---|---|
| 1 | TQD-Micromouse-JMⅡ | 1 | |
| 2 | 专用充电器 | 1 | |
| 3 | 专用下载器 | 1 | |
| 4 | 专用下载器连接线 | 1 | |
| 5 | USB线 | 1 | |
| 6 | 专用电池 | 1 | |
| 7 | 配套光盘 | 1 | |

# 附录D

## 教学内容和课时安排

本课程参考教学学时为60学时，具体分配表见表D-1。

表D-1 学时分配

| 序　号 | | 内　容 | 学　时 |
|---|---|---|---|
| 第一篇 | 基础知识篇 | 项目一　智能鼠的发展历程<br>项目二　智能鼠的硬件结构<br>项目三　智能鼠的开发环境<br>项目四　智能鼠的基本功能操作 | 22 |
| 第二篇 | 综合实践篇 | 项目一　智能鼠的运动姿态控制<br>项目二　智能鼠的智能控制算法和技术 | 18 |
| 第三篇 | 拓展竞技篇 | 项目一　智能鼠迷宫信息的获取与存储<br>项目二　智能鼠路径规划的原理<br>项目三　智能鼠走迷宫程序设计 | 20 |
| 总　计 | | | 60 |

附录E

电路图形符号对照表

电路图形符号对照表如表E-1所示。

表E-1　电路图形符号对照

| 序号 | 名称 | 国家标准的画法 | 软件中的画法 |
| --- | --- | --- | --- |
| 1 | 发光二极管 | | |
| 2 | 接地 | | |
| 3 | 按钮开关 | | |
| 4 | 二极管 | | |
| 5 | 绝缘栅场效应晶体管 | | |

# 附录F

## 专业词汇中英文对照表

专业词汇中英文对照表见表F-1~表F-3。

表F-1　与智能鼠相关的专业词汇

| 中文 | 英文 | 中文 | 英文 |
|---|---|---|---|
| 核心控制模块 | main control module | PWM信号发生器模块 | PWM signal generator driver module |
| 主控芯片 | main control chip | 左方、左斜、前方、右斜、右方 | the left, the left-oblique, the front, the right-oblique and the right |
| 输入模块 | input module | g段 | the g segment |
| 输出模块 | output module | 空心杯直流电动机 | coreless DC motor |
| 核心板电路 | main control circuit | 步进电动机 | stepping motor |
| 电源电路 | power circuit | 电动机驱动电路 | motor drive circuit |
| 控制电路 | control circuit | 真值表 | truth table |
| 外围电路 | peripheral circuit | H桥电路 | H-bridge circuit |
| 键盘显示电路 | keyboard-display circuit | 转动（步进电动机） | rotate |
| JTAG接口电路 | JTAG interface circuit | 电子元器件 | electronic component |
| 按键电路 | key-pressing circuit | 晶振 | crystal oscillator |
| 数据传输 | data transmission | 电容 | capacitance |
| 人机交互系统 | human-computer interaction system | 限流可调电阻 | adjustable current-limiting resistance |
| 吸地风扇技术 | suction fan technology | 数码管 | digitron |
| 占空比 | duty cycle | 外围器件 | peripheral devices |
| 角速度 | angular velocity | 脉冲振荡电路 | pulse oscillation circuit |
| 红外传感器 | IR sensor | 脉冲信号 | pulse signal |
| 红外检测电路 | infrared detection circuit | 方波 | square wave |
| 红外线 | infrared light | 感知系统 | perceptual system |
| 红外校准 | infrared calibration | 载波频率 | carrier frequency |
| 红外强度 | infrared intensity | 原理图 | schematic diagram |
| 红外发射头 | infrared transmitter | 软件界面 | software interface |
| 红外接收头 | infrared receiver | 驱动库 | driver library |

表F-2　与竞赛相关的专业词汇

| 中文 | 英文 |
| --- | --- |
| 单元格 | cell |
| 挡板 | wall |
| 立柱 | post |
| 竞赛场地 | competition maze |
| 起点 | the start |
| 目的地/终点 | the destination |
| 迷宫坐标 | the coordinate in the maze |
| 路口 | crossing |
| 电子自动计分系统 | electronic automatic scoring system |
| 参赛队员 | competitor |
| 智能鼠竞速比赛 | micromouse competition |
| 最优路径 | the optimal path |
| 轨迹 | trajectory |
| 通道 | passage way |

表F-3　与智能算法相关的专业词汇

| 中文 | 英文 | 中文 | 英文 |
| --- | --- | --- | --- |
| 底层驱动 | the bottom driver program | 差速控制 | differential-speed control |
| 顶层算法 | the top algorithm program | 直线运动 | straight movement |
| 算法 | algorithm | 转弯 | turning |
| 策略 | strategy | 校正车姿 | correct the attitude |
| 法则（左、右手法则） | rule | 运行校正 | attitude correction |
| 右手、左手、中心法则 | the right-hand rule, the left-hand rule, the central rule | 核心函数 | core function |
| 90°、180°转弯 | 90-degree turning/180-degree turning | （驱动步进电动机的）时序状态 | time sequence status |
| 编程并实现 | programming and realizing | 前进一格 | moving forward one cell |
| 等高图 | step map | 按键等待 | waiting for button press |
| 循环检测 | cycle detection | 判断车姿 | determining the attitude |
| "8字型"路径运行控制 | movement control in picture-8-shaped path | 暂停一步 | waiting one step |
| 实现避障 | obstacle avoidance | 精确转弯控制 | accurate turning control |
| 运动姿态的控制 | motion attitude control | 闭环控制 | closed-loop control |
| 两轮差速 | two-wheel difference speed | 绝对方向 | absolute direction |
| 路径规划和决策算法 | path planning and decision algorithm | 相对方向 | relative direction |
| 结构体 | struct | | |

# 附录G

## "智能鼠原理与制作"国际课程教学大纲

（适用于高等院校实训课程）

### 一、课程定位与特色

适用专业：电气类（0806）、电子信息类（0807）、自动化类（0808）、计算机类（0809）专业类中的基本专业与特设专业。

适用对象：服务于本科、应用型本科专业高年级，相关专业国际学生。

知识要点：模拟/数字电路、传感器、嵌入式硬件技术、嵌入式编程技术、自动化控制技术、高级智能算法设计等。

本课程定位与特色如下：

#### 1. 符合教育部《普通高等学校本科专业类教学质量国家标准》

本课程在通识类课程、专业基础类课程的学习基础上，为高年级学生的专业模块课程、工程设计、课程设计等理论和实践环节提供综合性教学载体，突出了"新工科"学科专业交叉融合的特点，突出了以学生为中心，注重激发学生的学习兴趣和潜能，创新形式、改革教法、强化实践，推动本科教学从"教得好"向"学得好"转变。

#### 2. 对接国际化教学理念，服务应用型人才培养

在符合我国标准的基础上，本课程内容与IEEE国际竞赛内容对接，在教材课程设置中体现"应用性、项目化、任务式"的设计理念，采取国际流行的任务导向、基于工作过程系统化、工程实践创新等思想，让学生在学习专业知识的同时，通过小组教学、沟通协作、分阶段目标等形式提升学生的全面综合素养。

#### 3. 服务"一带一路"，推广中国教育标准

本课程内容与多个国家的"鲁班工坊"建设项目高度融合。作为中国教育的新名片，本课程服务"一带一路"倡议，推广中国教育标准，为"一带一路"沿线国家提供丰富实践教学资源，服务各地技术技能人才培养。

## 二、课程目标

本课程以专业基础课程为前导，通过三个递进式模块和十个项目单元的学习训练，使学生具备智能机器人硬件设计与驱动、软件设计与编程、项目工程实施等工程素养；同时掌握嵌入式系统硬件设计、主流微处理器内部资源编程与配置、集成开发环境搭建，嵌入式设备人工智能编程与调试等知识技能；更深入理解传感器与检测信号调试、电动机精密控制、机器人智能搜索与路径规划等专业知识。

### 1. 知识目标

掌握嵌入式硬件系统的组成、嵌入式微处理器的特点、存储器结构；掌握传感器信号处理与分析方法；掌握空心杯直流电动机精确控制方法；掌握在迷宫中实现智能搜索与路径规划的相关知识。

### 2. 能力目标

掌握常用仪器仪表使用方法；掌握智能鼠设计、装配、调试的基本原理与方法；掌握嵌入式开发环境的搭建与基本使用方法；掌握对整体项目进行程序设计的方法；掌握智能鼠设备调试的常见技巧与故障分析与排除方法；具有独立进行资料与信息收集与整理的能力；具有制订与实施工作计划的能力；具备探究能力和运用理论知识解决实际问题的能力。

### 3. 素质目标

具有沟通及团队协作的能力；具有用于创新、敬业、乐业的工作作风和严谨、求精、求实的工匠精神；具有安全意识、质量意识与责任意识。

## 三、教学内容与要求

建议学时：60学时。

教学内容与要求见表G-1。

表G-1　教学内容与要求

| 序号 | 教学内容 | 教学要求 | 建议学时 | 重点及难点 |
|---|---|---|---|---|
| 1 | 第一篇　项目一　智能鼠的发展历程 | 了解智能鼠的起源，熟悉智能鼠的竞赛与调试环境 | 4 | 智能鼠的调试环境 |
| 2 | 第一篇　项目二　智能鼠的硬件结构 | 熟练掌握智能鼠硬件结构及各部分之间的相互关系；了解智能鼠关键器件的选型原则及方法；深刻理解智能鼠核心电路工作原理 | 6 | 智能鼠核心电路工作原理 |

| 序号 | 教学内容 | 教学要求 | 建议学时 | 重点及难点 |
|---|---|---|---|---|
| 3 | 第一篇 项目三 智能鼠的开发环境 | 熟悉IAR EWARM开发环境；熟练掌握智能鼠的程序下载技能 | 2 | IAR EWARM开发环境 |
| 4 | 第一篇 项目四 智能鼠的基本功能操作 | 理解智能鼠的红外检测和智能鼠的电动机驱动工作原理；熟练掌握智能鼠的姿态检测技术 | 8 | 姿态检测技术 |
| 5 | 第二篇 项目一 智能鼠的运动姿态控制 | 掌握使用PID算法稳定控制电动机转速的方法；掌握控制智能鼠转弯的编程实现；熟练掌握智能鼠转弯角度的精确控制的调试方法 | 8 | PID算法的原理与实现 |
| 6 | 第二篇 项目二 智能鼠的智能控制算法和技术 | 掌握控制智能鼠运行在跑道正中并避免碰撞前方挡板的方法；掌握智能鼠识别和记忆自身坐标的方法；掌握智能鼠的吸盘控制和形状设计的方法 | 8 | 智能鼠居中行走、地图坐标表示、挡板信息记录 |
| 7 | 第三篇 项目一 智能鼠迷宫信息的获取与存储 | 掌握相对方向与绝对方向转换的原理与实现；掌握常用的地图搜索策略的编程实现方法 | 8 | 迷宫搜索优化法则的原理 |
| 8 | 第三篇 项目二 智能鼠路径规划的原理 | 掌握迷宫搜索优化法则的原理及实现方法；掌握等高图的原理与使用方法 | 8 | 相对方向与绝对方向的转换 |
| 9 | 第三篇 项目三 智能鼠走迷宫程序设计 | 理解智能鼠竞赛基本程序的结构和工作流程 | 8 | 根据不同竞赛策略调整程序结构 |

## 四、实践教学

第一篇 项目二 智能鼠PCB设计（2学时）

（1）目的：学习智能鼠的硬件设计。

（2）内容：根据硬件结构图进行设备选型和PCB设计。

（3）步骤：①设备选型；②确定元器件走线需求；③设计智能鼠尺寸外观；④PCB的设计和绘制。

（4）分组：3~5人一组。

（5）仪器设备：计算机、Altium软件、关键元器件数据手册。

第一篇 项目三 开发环境的搭建与使用（2学时）

（1）目的：学习开发环境的搭建与使用。

（2）内容：安装、设置并熟悉开发环境。

（3）步骤：①安装IAR开发环境；②配置IAR开发环境；③熟悉IAR开发

环境的基本操作。

（4）分组：独立完成。

（5）仪器设备：计算机、集成开发环境。

第一篇　项目四　智能鼠的基本功能控制（6学时）

（1）目的：学习控制智能鼠完成一些基本动作。

（2）内容：红外检测、电动机驱动和姿态检测。

（3）步骤：①编写红外驱动和传感器信息采集程序；②编写直流电动机驱动程序；③编写角速度传感器信息采集程序。

（4）分组：3~5人一组。

（5）仪器设备：计算机、集成开发环境、智能鼠、LED调试板。

第二篇　项目一　智能鼠的运动姿态控制（6学时）

（1）目的：学习精确控制电动机运转，并使用角速度传感器精确控制智能鼠姿态。

（2）内容：智能鼠稳定直行和精确转弯。

（3）步骤：①编写程序，使用PID算法稳定控制电动机转速；②编写程序精确控制智能鼠转弯角度；③编写程序控制智能鼠在行进中精确转弯。

（4）分组：3~5人一组。

（5）仪器设备：计算机、集成开发环境、智能鼠、3×3迷宫。

第二篇　项目二　智能鼠的智能控制算法和技术（6学时）

（1）目的：学习控制智能鼠自动避障、更新和记忆坐标信息，掌握吸地风扇的智能化操控。

（2）内容：智能鼠自动避障、更新坐标、记忆坐标信息、利用风扇提高稳定性。

（3）步骤：①编写程序利用红外传感器控制智能鼠运行在跑道正中；②编写程序控制智能鼠识别前方障碍并及时制动；③编写程序控制智能鼠精确前进若干个方格并记忆自己的坐标；④编写程序控制吸盘转速。

（4）分组：3~5人一组。

（5）仪器设备：计算机、集成开发环境、智能鼠、16×16迷宫。

第三篇　项目一　地图信息的获取与存储（6学时）

（1）目的：学习控制智能鼠自动避障和记忆地图信息。

（2）内容：智能鼠自动避障、识别地图信息、记忆地图信息。

（3）步骤：①编写程序实现相对方向与绝对方向的相互转换；②编写程

序利用方向的转换和红外传感器的信息获取正确的地图信息；③编写程序实现常见的地图搜索策略。

（4）分组：3~5人一组。

（5）仪器设备：计算机、集成开发环境、智能鼠、16×16迷宫。

第三篇　项目二　智能鼠的路径规划原理和行为决策法则（6学时）

（1）目的：学习智能鼠进行智能化的地图搜索并找到通往终点的最佳路径。

（2）内容：智能鼠常见搜索策略，路径寻优的常见算法。

（3）步骤：①编程实现常用的搜索策略；②编写程序利用Flood扩散算法优化搜索程序；③编写程序利用等高图算法实现冲刺程序。

（4）分组：3~5人一组。

（5）仪器设备：计算机、集成开发环境、智能鼠、16×16迷宫。

第三篇　项目三　智能鼠走迷宫竞赛（6学时）

（1）目的：理解智能鼠竞赛基本程序的结构和工作流程。

（2）内容：系统学习智能鼠搜索、首次冲刺、全图搜索、二次冲刺等程序。

（3）步骤：编写程序将智能鼠搜索、首次冲刺、全图搜索、二次冲刺合理串联，并利用传感器结合地图特点合理配置智能鼠工作流程。

（4）分组：3~5人一组。

（5）仪器设备：计算机、集成开发环境、智能鼠、16×16迷宫。

## 五、教学方法建议

（1）教师应依据工作任务中的典型产品为载体安排和组织教学活动。

（2）教师应按照项目的学习目标编制项目任务书。项目任务书应明确教师讲授（或演示）的内容；明确学习者预习的要求；提出该项目整体安排以及各模块训练的时间、内容等。如以小组形式进行学习，对分组安排及小组讨论（或操作）的要求，也应做出明确规定。

（3）教师应以学习者为主体设计教学结构，营造民主、和谐的教学氛围，激发学习者参与教学活动，提高学习者学习积极性，增强学习者学习信心与成就感。

（4）教师应指导学习者完整地完成项目，并将有关知识、技能与职业道德和情感态度有机融合。

## 六、考核及成绩评定方式

（1）考核方式：考查。

（2）成绩评定办法：理论和实际操作考核。

总评成绩以百分制计算，分为平时成绩考核和期末综合考核两部分。平时成绩考核一般为出勤、项目作业、测验，占总成绩的 30%；期末综合考核一般为理论考试、实际操作考核，占总成绩的 70%，其中理论考试占总成绩的30%，实际操作考核占总成绩的40%。

## 七、与其他课程的联系

### 1. 先修课程及联系

先修课程主要有：计算机原理与组成、电子技术基础、信号处理、传感器技术与应用、微处理器原理与应用、嵌入式系统与设计、人工智能原理与应用等。

### 2. 后续课程

毕业设计。

## 八、教材及参考资料

### 1. 推荐教材：

王超, 高艺, 宋立红. 智能鼠原理与制作：高级篇[M]. 北京：中国铁道出版社有限公司，2019.

### 2. 参考资料：

[1] 沈健良，贾玉坤，周芬芬，等．STM32F10X系列之ARM微控制器入门与提高[M]．北京：北京航空航天大学出版社，2013.

[2] 黄智伟，王兵，朱卫华．STM32F 32位ARM微控制器应用设计与实践[M]．北京：北京航空航天大学出版社，2012.

## 九、编写与审核

编写人：邓蓓　宋立红　高源　董瑞　陈立考　邱建国
审核人：龚威　王超　高艺
2020年6月26日

Intelligent Micro Motion Device (Micromouse) Technology and Application Series
Research Result of Tianjin "the Belt and Road" Joint Laboratory (Research Center)
Engineering Practice Innovation Project (EPIP) Teaching Mode Planned Textbook

# Micromouse Design Principles and Production Process
## (Advanced)

**Compiled by** Wang Chao, Gao Yi, Song Lihong
**Translated by** Zhou Fanyu, Yan Jingyi, Liu Jia

中国铁道出版社有限公司
CHINA RAILWAY PUBLISHING HOUSE CO., LTD.

# Introduction to the contents

The book is bilingual in both Chinese and English, based on the TQD-Micromouse-JQ provided by Tianjin Qicheng Science and Technology Co., Ltd., which is the improve-level of a series of books on Micromouse Technology and Application.

The book is based on real engineering projects, through "Elementary Knowledge", "Comprehensive Practice" and "Advanced Skills and Competitions". All of these three chapters are describe the development, hardware, development environment, and function debugging of the Micromouse; basic operation of Micromouse, the movement posture control, intelligent control algorithm and technology of Micromouse, etc. The appendix of this book provides the relevant knowledge of the international Micromouse maze competition, such as the device list of TQD-Micromouse-JQ, the Micromouse maze library, bilingual comparison table of glossary, and the international curriculum standard.

The book is equipped with a wealth of resources such as videos, pictures, texts, etc. on important knowledge points, skills points and literacy points. The reader can obtain relevant information by scanning the QR code in the book.

The book is suitable for colleges and universities, especially for "new engineering" major, as a guidance book of teaching information and automation technology integration and innovation course. It can also be used as a training book for relevant engineering and technical personnel or reference book for Micromouse lovers.

**About the Authors**

### Wang Chao

Wang Chao is currently a Professor in School of Electrical and Information Engineering, Tianjin University, China. He is a member of the Teaching Guiding Committee for Automation Majors under the Ministry of Education of China. His current research interests include multiphase flow measurement and instrumentation, electrical tomography (ERT, ECT, EMT and EST). His courses at Tianjin University include computer control technology and industrial control networks. Since 2010, he has introduced Micromouse as an important carrier of practical teaching into school of electrical and information engineering for the first time. Two teams of Tianjin University won the first and second place at 2018 APEC Micromouse competition.

### Gao Yi

Gao Yi is a master supervisor of School of Electronic Information Engineering at Nankai University, deputy director of Electronic Information Experimental Teaching Center, deputy director of Youth Working Committee of Tianjin MCU Society and member of the judging group of many technology competition of college students and Tianjin vocational skills competition. He has participated in "national high-tech research and development plan (863 plan) projects", "Tianjin science supporting plan key projects" and multiple horizontal scientific research projects. He has led teams to participate in the national undergraduate electronic design competition, Tianjin electronic design competition, Tianjin internet of things competition, Tianjin IEEE Micromouse competition for college students, APEC international Micromouse competition and national robot competition.

### Song Lihong

Song Lihong is general manager of Tianjin Qicheng Science and Technology Co., Ltd. and founder of Qicheng Micromouse. The company has been committed to the research and development, design, production, promotion and service work of teaching instruments about embedded system, internet of things and artificial intelligence used in higher education, vocational education and basic education. Qicheng has sponsored the "Qicheng Cup" Micomouse maze competition of college students and the intelligent micro motion device (Micromouse) competition of vocational college skills competition more than 40 times. Since 2016, the company has been actively engaged in the technical support service of international project Luban Workshop. As an innovative educational equipment in China, Qicheng Micromouse has been brought to Thailand, India, Indonesia, Pakistan, Cambodia, Nigeria, Egypt and other countries and has been favored by teachers and students in these countries. Qicheng Micromouse has made contributions to "the Belt and Road" initiative.

## Zhou Fanyu

Zhou Fanyu, master of English education, University of Central Oklahoma, is currently working in the international exchange and cooperation office of Tianjin Vocational College of Mechanics and Electricity. During her stay in the United States, she participated in research projects such as "the Comparative Study of Chinese and English Languages". She taught a large number of non-American students whose English was the second language in Confucius Classroom, and gained a lot of teaching experience. In 2018, she participated in the summer Davos forum and served as the interpreter and affairs officer for Sun Xiansheng, secretary general of the International Energy Forum. In 2018, she participated in the establishment and construction of Portugal Luban Workshop. In 2018, she published a bilingual paper on "Research on the Training of Practical Ability in Micromouse Project by the Teaching Model of Luban Workshop". In 2019, she participated in the establishment and construction of Madagascar Luban Workshop. During this period, she completed the translation of several meetings, accompanying translation, and translated more than 200,000 words of documents and materials.

## Yan Jingyi

Yan Jingyi, assistant to the general manager of Tianjin Qicheng Science and Technology Co.,Ltd., studied in the University of California, Santa Cruz. In 2015, she served as the accompaniment interpreting for professor MIT David Otten, chairman of the organizing committee of APEC Micromouse international competition. In 2018, she served as the English translator and simultaneous interpreter for Mr. Davaanyam, vice chairman of the board of directors of new Mongolia Education Group when he visited in Tianjin. In 2018, she went to Cambodia as a volunteer interpreter for Bun Phearin, president of National Polytechnic Institute of Cambodia. She was helping Cambodia students learn the Chinese Micromouse technology at the same time. Yan makes contributions to the education in countries along "the Belt and Road" for a long time.

## Liu Jia

lecturer of General English Teaching department of Nankai University as well as PhD student of foreign language department, Nankai university. She is the lecturer of two national excellent courses,"College English" and "Scientific Research Methodology". She has been twice rated as an excellent teacher of General English Teaching department of Nankai University. She has won the first prize of Nankai University teaching competition and the second prize of Tianjin basic teaching skills competition. She has led and participated in several municipal and university level scientific research projects, and published dozens of research papers. She has compiled oral English textbooks, translation qualification examination textbooks, writing textbooks and dictionaries, and translated several English books.

Micromouse is a micro, intelligent motion device (or embedded microrobot) composed of embedded micro controllers, sensors and electromechanical moving parts. Micromouse can reach the predetermined destination fast by automatically memorizing all the routes in various mazes and selecting the optimal path out with the aid of suitable algorithms. The Micromouse competition involves wide-ranging scientific knowledge such as mechatronics, cybernetics, optics, programming and artificial intelligence.

For more than 40 years, the Institute of Electrical and Electronics Engineers (IEEE) has hosted an annual Micromouse competition. Since its inception, the international event has featured extensive and active participation, especially that of students from colleges and universities in the US and Europe. Some universities even offer an elective course on the principles and production process of Micromouse. In 2007, Shanghai and some other cities in the Yangtze River Delta started to stage small-scale, experimental Micromouse competitions. In 2009, Tianjin Qicheng Science and Technology Co., Ltd. introduced the competition to Tianjin and added local features to create an updated version based upon the Engineering Practice Innovation Project (EPIP) teaching model. These pioneering efforts helped boost future Micromouse events and played a key role in integrating them into classroom teaching. Years of exploration and progress have made Micromouse competitions to serve as educational platforms that encourage innovation and practice. By bringing together expertise and interest, the multi-dimensional and pioneering competitions have been essential for cultivating students' capabilities in practice and innovation, reforming curriculum and improving education.

In order to further promote and apply the achievements of Micromouse, we specially organized the book of *Miromouse Design Principles and Production*

*Process* (*Advanced*) for undergraduates. This book is based on the TQD-Micromouse-JM II provided by Tianjin Qicheng Science and Technology Co.,Ltd. As a teaching carrier, practice teaching is carried out from shallow to deep, and from easy to difficult, step-by-step teaching.

This book will teach users with basic principles of Micromouse and end up as a profession in the field by following a step-by-step method that is also used in compiling the book. After reading the book, users will acquire more knowledge about engineering practice, enrich their experience in technology application, open up a broader vision in expertise, and become more professional.This book's ultimate goal is to cultivate innovation-minded practitioners. The select cases in the book are all adapted from real-life engineering projects. Also all the authors are came from enterprises, colleges and universities that have long remained committed to R&D on Micromouse or that have won awards in international competitions.

This book offers abundant videos, pictures and texts, etc. on important knowledge Points, skills points and literacy points. The reader can scan the QR code in the book to get these supporting resources. The authors' rich international teaching experience has made the book became a practical teaching carrier to promote international talent training. Intelligent micro motion device (Micromouse) technology and application series are the research Result of Tianjin "the Belt and Road" Joint Laboratory (Research Center)—Tianjin Sino-German and Cambodia Intelligent Motion Device and Communication Technology Promotion Center and are also the Engineering Practice Innovation Project (EPIP) teaching model planning materials. The book is suitable for colleges and universities, especially for "new engineering" major, as a guidance book of teaching information and automation technology integration and innovation course. It can also be used as a training book for relevant engineering and technical personnel or reference book for Micromouse lovers.

The syllabus of the international course is listed in the appendix of

"Micromouse Design Principles and Production Process" (for undergraduates). Taking the professional basic courses as the introduction, instruction from three progressive modules and ten projects will ensure students' engineering literacy, such as hardware design and drive and software design and programming of Micromouse and project implementation. The students will also have an in-depth understanding of sensors and detection signal debugging motor precision control, Micromouse intelligent search and path planning and other professional knowledge. The course content is highly integrated with the Luban Workshop construction projects in many countries. Serving "the Belt and Road" initiative, spreading Chinese vocational education standards, providing practical teaching resources for all countries along "the Belt and Road" route, Serving the training of skilled personnel in various fields.

This book is co-authored by Wang Chao, professor of Tianjin University; Gao Yi, associate professor of Nankai University; and Song Lihong, general manager of Tianjin Qicheng Science and Technology Co.,Ltd. the founder of Qicheng Micromouse. The English version was translated and compiled by Zhou Fanyu, teacher in Tianjin Vocational College of Mechanics and Electricity, Yan Jingyi, a general manager assistant of Tianjin Qicheng Science and Technology Co., Ltd. and Liu Jia, lecturer of General English department of Nankai University. Associate Professor Zhang Lian and lecturer Gao Yuan, from Tianjin Sino-German University of Applied Sciences participated in the collation and translation of this book. David Otten, professor of Massachusetts Institute of Technology (MIT) in USA, Peter Harrison, professor of Birmingham City University in UK and António Valente, professor of University of Trás-os-Montes and Alto Douro in Portugal are proofreaders of the English version and they specially wrote congratulatory letters for this books. The book has received great and generous support from scholars and experts who come from Tianjin University, Nankai University, Tianjin Sino-German University of Applied Sciences, Massachusetts Institute of Technology (MIT), Birmingham City University, and University of Trás-os-Montes and Alto Douro. Chen Likao,

Qiu Jianguo and Song Shan, are the employees of Tianjin Qicheng Science and Technology Co.,Ltd. provided practical engineering cases, QR codes, videos and PPT course resources for the book. We also owe great gratitude to Tianjin Municipal Education Commission, China Railway Publishing House Co., Ltd. and Tianjin Qicheng Science and Technology Co.,Ltd. for their invaluable guidance and support. The bilingual version of this book is sponsored compiled by Tianjin Light Industry Vocational Technical College, published by China Railway Publishing House Co., Ltd. and will be used in countries along the "Belt and Road" route through the Luban Workshop program.

There will be some gaps or even errors in the book due to a tight publishing schedule and insufficient consideration from the authors, so any constructive criticisms and suggestions are greatly welcomed.

Authors

August, 2020

# Contents

# Chapter 1

# Elementary Knowledge

  The Micromouse competitions have enjoyed worldwide popularity for over four decades. Micromouse is required to search the entire maze without human manipulation to find the destination. And then, Micromouse needs to select, among the many possible paths, the optimal path to reach the destination and spurt from the start to the destination as quickly as possible. Contestants are ranked by the search time plus the spurt time of Micromouse. Mazes used in competitions comply with the international standards set by the Institute of Electrical and Electronics Engineers (IEEE). In this chapter, you will gain a systematic understanding of Micromouse technology from international standard mazes of IEEE, hardware systems and software development environment. You will also learn in more detail the fundamental principles and practical operations of Micromouse.

# Project 1

## Evolution of Micromouse

 Learning objectives

(1) Learning about the evolution of Micromouse.

(2) Understanding the Micromouse competition platform, i.e. mazes and automatic scoring system.

# Task 1　Origins of Micromouse

### 1. Birth of Micromouse

In 1938, Claude Elwood Shannon, an American mathematician born in Michigan state, completed his master's thesis *A Symbolic Analysis of Relay and Switching Circuits*. He used the Boolean algebra that happens to correspond with the binary system of 0 and 1 to process the relay switches of information in a pulse mode. The notable work had transformed the design of digital circuits both theoretically and technologically, making it an epoch-making thesis in the modern history of digital computers.

In 1948, Shannon published another famous work that is still relevant today, *A Mathematical Theory of Communication*, which earned him the title "Father of Information Theory".

In 1956, Shannon attended the Dartmouth Conference and became one of the founding fathers of the emerging discipline of artificial intelligence. He pioneered the application of artificial intelligence in computer chess and invented a mechanical mouse that could run through a maze autonomously, which proved that computers could improve their intelligence through learning.

## 2. Evolution of Micromouse in the World

In 1972, *Journal of Mechanical Design* started a contest where mechanical mouse solely driven by mousetrap springs competed with other entries to see which one could cover the longest distance.

In 1977, *IEEE Spectrum* introduced the concept of Micromouse, which is a small robotic vehicle controlled by microprocessors and has the capabilities to decode and navigate in complex mazes.

In 1979, the IEEE initiated a Micromouse competition through its magazine (*Spectrum and Computer*) and it rewarded the designer of the champion Micromouse that could find a way out of the maze all on its own in the shortest possible time span with USD 1,000.

In 1980, the first All Japan Micromouse International Competition was held, followed with more such events, such as UK Micromouse competition in 1980, Singapore IES Micromouse Competition in 1987, and Micromouse Competition for College Students held by China Computer Federation (CCF) in 2007. As shown in Fig.1–1–1.

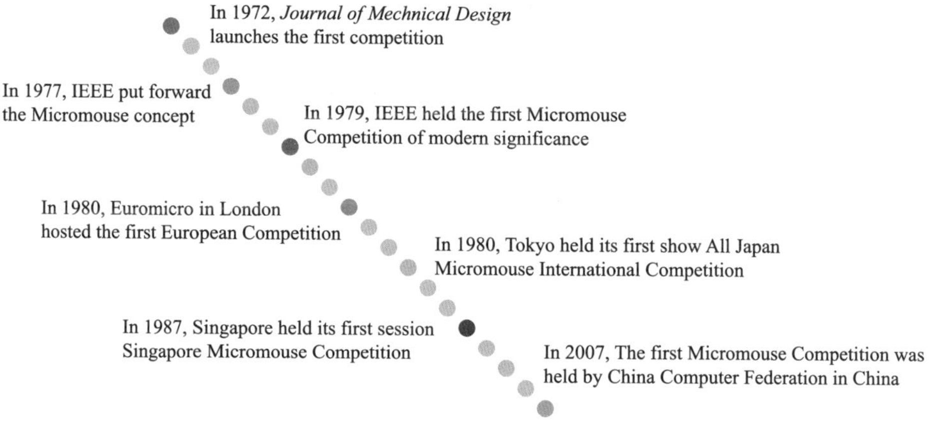

In 1972, *Journal of Mechnical Design* launches the first competition

In 1977, IEEE put forward the Micromouse concept

In 1979, IEEE held the first Micromouse Competition of modern significance

In 1980, Euromicro in London hosted the first European Competition

In 1980, Tokyo held its first show All Japan Micromouse International Competition

In 1987, Singapore held its first session Singapore Micromouse Competition

In 2007, The first Micromouse Competition was held by China Computer Federation in China

Fig. 1–1–1　Global development trajectory

The past four decades have witnessed the great evolution of Micromouse from the mechanical mouse in 1972 to Micromouse nowadays. The competitions now feature wider participation at all education levels from around the globe. When the competitions were first launched, only graduate students from world-renowned colleges and universities such as Harvard and MIT were able to participate. Later on, students from research universities, universities of applied

sciences and vocational schools could compete as well. And nowadays even primary and middle school students may take part in Micromouse competitions. Micromouse has been adopted as a teaching vehicle at educational institutions of various levels to cultivate students' engineering literacy, improve the awareness of innovation and boost their design skills.

Micromouse competitions in various forms are flourishing across the world. Now they have grown into global innovation events that are applicable to students at different education levels.

## 3. Evolution of Micromouse in China

Micromouse competitions have experienced over ten years of growth since 2007 in China, as shown in Fig.1–1–2. In 2007, Tianjin Qicheng Science and Technology Co., Ltd. first introduced the competition to Tianjin and added local features to create an updated version with the advanced Engineering Practice Innovation Program model as a core vision. These pioneering efforts helped boost future Micromouse events in China and played a key role in integrating relevant technologies into classroom teaching.

Video

Evolution of Micromouse in China

| First Competition | Higher Education | Vocational Education | General Education | International Competition |
|---|---|---|---|---|
| In 2007, The First Micromouse Competition was held in China | Form 2009, Micromouse Competition has always been a university students' subject competition | Form 2010, Micromouse Competition has always been a colleges students' skills competition | Form 2016, General Education and vocational Education Micromouse Challenge Contest started to be held in China | Form 2016, International Invitational Micromouse Competition is held every year in China |

Fig. 1–1–2   Development trajectory in China

The competitions have helped upgrade industries, broaden the horizon, accumulate experience in practice and innovation and cultivate high-caliber, high-tech and highly skilled personnel (see Fig.1–1–3). A variety of Micromouse competitions has been held in China, such as contests for university students, competitions for vocational colleges, and international challenge competitions for general-vocational high schools, which enable us to gather rich experience and solid technical prowess.

The past decade has witnessed China constantly exploring new ways to make the local Micromouse competitions gain increasing international exposure.

When the competitions were first introduced into China, we simply copied foreign models, but as we gain more experience and build novel platforms for international exchanges and cooperation, foreign countries are also learning from us. Generally speaking, there are three stages in the development of Micromouse in China: imitation and learning; innovation and growth; and going global as a leader.

Fig. 1–1–3　Students at Micromouse competitions

The first stage: In 2015, the Tianjin team went to the United States to participate in the 30th APEC Micromouse Contest and ranked sixth globally (see Fig.1–1–4). In 2017 and 2018, Tianjin Qicheng Technology sponsored the champion team that won the enterprise designated-topic session at Tianjin College Students Micromouse Competition to go to Tokyo and compete in the 38th and 39th All Japan Micromouse International Competition (see Fig.1–1–5). The travel and boarding fees of the contestants were fully covered by the company. The two competitions in Japan enabled us to learn more about global advanced technologies of Micromouse and make connections with industry experts and leaders, greatly boosting the development of Micromouse technologies in China.

● Video

Imitation and learning

Fig. 1–1–4　Tianjin team in the US for APEC 30th Annual Micromouse Contest

Fig. 1-1-5    Tianjin team at All Japan Micromouse International Competition

The second stage: Micromouse competitions were added local, innovative features and underwent necessary reforms to comply with China's realities. A wide range of tiered teaching platforms based upon the TQD-Micromouse produced by Tianjin Qicheng Technology were created to meet the needs of students at different levels: junior high, senior high, undergraduate, and postgraduate. Since 2016, IEEE Micromouse International Invitational Competition in China has featured more extensive participation by world-famous scholars and experts, domestic and foreign teachers and students, and elite teams in China. The name list includes Professor David Otten from MIT, Professor Su Jinghui from Lunghwa University of Science and Technology in Taiwan, China, Professor Huang Mingji from Ngee Ann Polytechnic in Singapore, Professor Peter Harrison from Birmingham City University, and Mr. Yoko Nakagawa, Secretary-General of the Organizing Committee of All Japan Micromouse International Competition; faculty and students from Luban Workshop in Thailand, India, Indonesia, Pakistan and Cambodia; and competition teams from Tianjin, Beijing, Henan and Hebei, to name a few (see Fig.1-1-6 and Fig.1-1-7). By signing up for the competitions held in China, international contestants learned more about the Chinese standards, rules, models and philosophy and later accepted them. In this way, global exchanges and collaboration was facilitated and both sides had something meaningful to learn from each other.

The third stage: Educational opening-up is integral to China's reform and opening-up initiative. As "the Belt and Road" initiative gains momentum, the Luban Workshop programme has been launched since 2016 under the guidance of the Ministry of Education. Micromouse, an exemplar of China's excellent teaching aid, has gone global thanks to the programme. Since then,

Video

Innovation and growth

● Video

Going global
as a leader

Tianjin Qicheng Technology has gone to a raft of foreign countries such as Thailand, India, Indonesia, Pakistan, Cambodia, Nigeria and Egypt to promote Micromouse competitions and offer training sessions free of charge, which are well-received by both the local teachers and students (see Fig.1–1–8-Fig.1–1–13). Micromouse has served as a bridge connecting China with the rest of the world!

Fig. 1–1–6    2018 Third IEEE Micromouse International Invitation

Fig. 1–1–7    2019 "Qicheng Cup" IEEE Micromouse International Invitation

Fig. 1–1–8    Micromouse training session at Luban Workshop in India

Fig. 1-1-9　Micromouse training session at Luban Workshop in Thailand in 2016

Fig. 1-1-10　Micromouse training session at Luban Workshop in Indonesia in 2017

Fig. 1-1-11　Micromouse training session at Luban Workshop in Pakistan in 2018

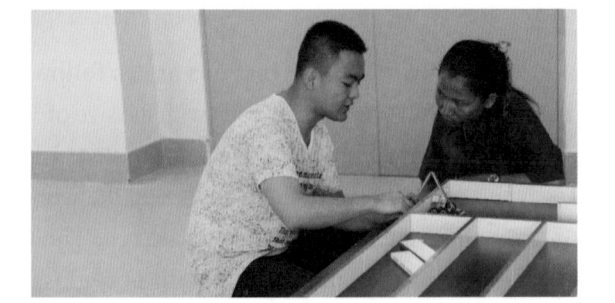

Fig. 1-1-12　Micromouse training session at Luban Workshop in Cambodia in 2018

Fig. 1–1–13  Micromouse training session at Luban Workshop in Egypt in 2020

# Task 2  Competition and Debugging Environment of Micromouse

## 1. Competition maze

At present, the international Micromouse competition adopts IEEE standard and uses the same specification maze, that is, a square maze composed of 16×16 cells. The "walls" of the maze can be inserted, so that a variety of mazes can be formed.

The TQD-Micromouse Maze 16×16 is shown in Fig.1–1–14. The floor of the maze is 2.96 m×2.96 m, and there are 16×16 standard maze cells on it. Wall and post of the classical Micromouse maze is shown as Fig.1–1–15.

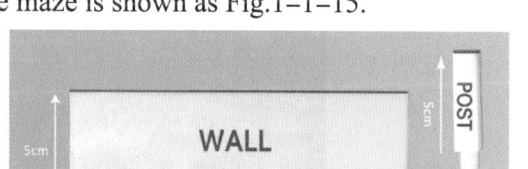

Fig. 1–1–14  TQD-Micromouse Maze 16×16

Fig. 1–1–15  Wall and post of the classical Micromouse maze

TQD-Micromouse maze 16×16 specifications are as follows:

(1) The maze is composed of 16×16 square cells with the size of 18 cm×18 cm.

(2) The height of the walls are 5 cm and their thickness are 1.2 cm, so the actual distance of the passageways are 16.8 cm, and the walls seal the whole maze.

(3) The side of the walls are white, and the top are red. The floor of the maze is painted in black. It is made of wood, finished with non-gloss black paint. The paint on the side and top of the wall can reflect infrared light, and the floor can

absorb infrared light.

(4) The start can be set at one of the four corners. The start must have three walls and only one exit. The destination is located at the center of the maze, which is composed of four cells.

(5) There are small posts, each 1.2 cm×1.2 cm×5 cm, can be inserted at the four corners of each cell. The position of the posts are called lattice points. There are at least one wall to a lattice point except for the destination.

(6) The dimensional accuracy error of the maze making should be no larger than 5%, or less than 2 cm. The joint of the maze floors shall not be more than 0.5 cm, and the gradient change of the joint point shall not be more than 4°. The gap between the walls and posts shall not be more than 1 mm.

(7) The start and the destination shall be designed based on IEEE Micromouse competition rules and standards, that is, Micromouse starts in a clockwise direction.

## 2. Special testing site

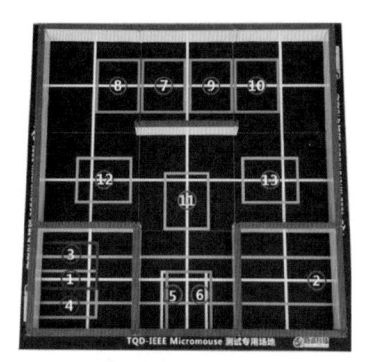

Fig. 1-1-16   Special testing site for TQD-IEEE Micromouse

There are 13 marked positions painted on the special testing site and different colors are used to distinguish them (see Fig.1-1-16). They are used for aiding adjusting infrared and turning parameters. Next, let's to learn them:

(1) ① to ②, gray passageways, which is used to detect the offset of Micromouse in the absence of infrared calibration.

(2) ③ dark red rectangle, ④ orange rectangle; ③ to ②, ④ to ② are both used to check Micromouse's forward going condition with infrared calibration.

(3) ⑤ yellow rectangle is used to adjust the left front infrared intensity of Micromouse, ⑥ green rectangle is used to adjust the right front infrared intensity of Micromouse ; Correct the attitude.

(4) ⑦, ⑧ green rectangles are used to adjust the right rear infrared intensity of Micromouse , ⑨, ⑩ green rectangles are used to adjust the left rear infrared intensity of Micromouse ; Detect the intersection.

(5) ⑪、 ⑫、 ⑬three blue rectangles are used to debugging 90-dcgree turning of Micromouse.

## 3. Automatic scoring system

In order to accurately measure the time used by Micromouse to complete

● Video

The working principle of automatic scoring system ●

the competition, it is necessary to calculate the time of Micromouse passing the start and the destination full-automatically. The electronic automatic scoring system designed and produced by Tianjin Qicheng Science and Technology Co., Ltd., which is specially used for Micromouse competition, is shown in Fig.1–1–17.

TQD-Micromouse Timer V2.0 system includes the start infrared detection module, the destination infrared detection module, the scoring system module, and the scoring software, etc.

The start infrared detection module and the destination infrared detection module are charged through mini USB. Through a set of inner-placed thru-laser sensors, it can detect the passing of Micromouse. The scoring system module is used to receive the data sent by the infrared detection modules through ZigBee. After been processed by the scoring software in computer, the running condition of Micromouse in the maze is shown in a visualized manner. The scoring software can also be used alone. The start event and destination event can be input through mouse. The overall timing accuracy of the scoring system can reach 0.001 s.

Fig. 1–1–17　TQD-Micromouse Timer V2.0

Fig.1–1–18　The start

The start infrared detection module and the destination infrared detection module are respectively installed in the start cell and the destination cell, as shown in Fig.1–1–18, Fig.1–1–19. When Micromouse passes by, laser beam is blocked, thus generates a start or destination signal.

Fig.1–1–19　The destination

## Reflection and Summary

(1) What are the components of the Micromouse maze based on IEEE standard?

(2) What are the characteristics of Micromouse competition?

(3) The usage automatic scoring system has improved the accuracy of competition result calculation greatly. Please briefly explain it's working principle.

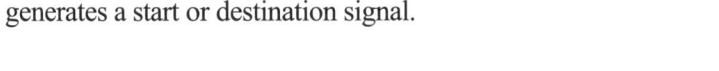

# Project 2

## Micromouse Hardware Structure

 Learning objectives

(1) Understanding the basic hardware structure of Micromouse.

(2) Mastering the application of Micromouse CPU.

TQD-Micromouse-JM II (see Fig. 1-2-1) is an intelligent robot device designed and produced by Tianjin Qicheng Science and Technology Co., Ltd. The maximum design speed is 5 m/s, which can be used in competitions. At the same time, it can also be used as a mobile robot development and research platform, teaching and scientific works exhibition, performance and other fields. It is a model of artificial intelligence robots.

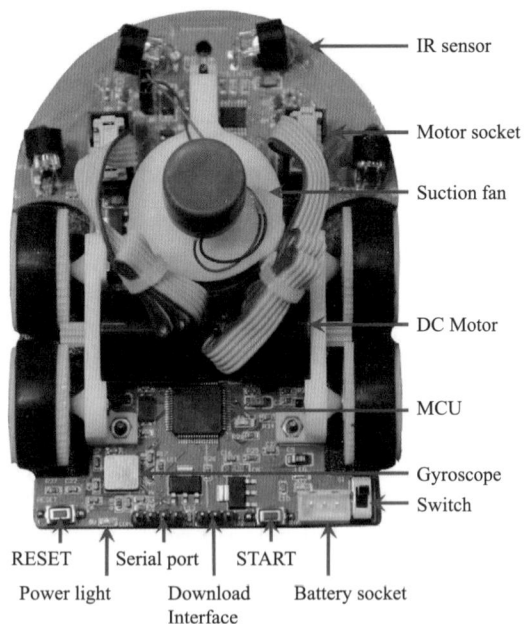

Fig. 1-2-1    TQD-Micromouse-JM II hardware layout

# Task 1   Compositions of Micromouse

(1) The core controller uses high-performance STM32 chip, 72 MHz operating frequency, 512 KB Flash, 64 KB RAM, powerful, highly integrated, and efficiently handle all kinds of events in the operation of Micromouse.

(2) High-performance low-power FAULHABER1717SR Germany coreless DC motor, maximum speed 14,000 r/min, built-in industrial-grade integrated 1024 line encoder, easy to achieve high-precision control of Micromouse speed and position.

(3) The ADXRS620 gyroscope detects the offset angle of Micromouse, achieves high-precision control of Micromouse running posture and turning angle.

(4) The motor drive adopts DRV8848 chip, dual H-bridge design, and cooperates with PWM technology to effectively respond to the motor's starting, braking, positive and negative rotation control.

(5) Incorporating TQD suction fan technology, the suction power can reach 0.5kg, enhance the grip, effectively overcome the tire slip, and achieve high-speed turning.

(6) Four sets of analog infrared sensors can accurately measure the distance from the labyrinth baffle.

(7) The power source is powered by 7.4 V high-quality lithium battery, with excellent voltage-stabilizing circuit design, and equipped with anti-reverse insertion and charging reminder functions to ensure the safe and stable operation of Micromouse.

(8) The mechanical structure is 3D printed with high-strength resin. It has light weight, high strength and high precision, which makes the Micromouse body firm, lowers the center of gravity, enhances grip, and avoids the "speeding" phenomenon during high-speed operation.

(9) This Micromouse adopts IAR as the software development environment and provides open source DEMO program package, including infrared automatic detection code and 45° walking code.

(10) The combination of the central algorithm and the flood algorithm makes the maze solution more efficient and gets rid of the problem that the complex maze is difficult to find the end point in the past.

The block diagram of the circuit composition of TQD-Micromouse-JM Ⅱ is shown in Fig. 1–2–2, which is mainly composed of the main controller, infrared transmitting and receiving, motor driving, suction motor, encoder and gyroscope, etc.

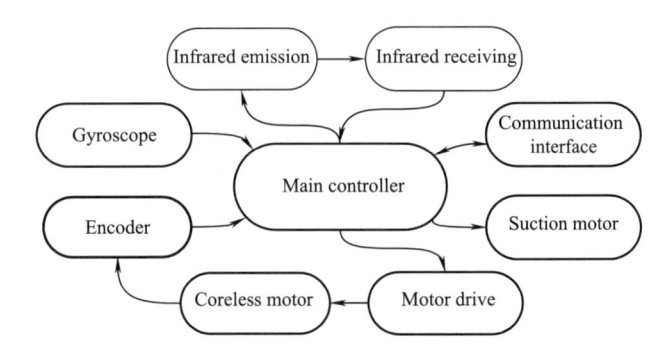

Fig. 1–2–2   TQD-Micromouse-JM Ⅱ circuit composition block diagram

# Task 2   The Core Control Circuit of Micromouse

TQD-Micromouse-JM II uses the PWM generator of STM32F103RET6, GPIO port, SPI interface, counter/timer, A/D conversion module, serial port and I²C interface. Its minimum system principle is shown in Fig. 1–2–3.

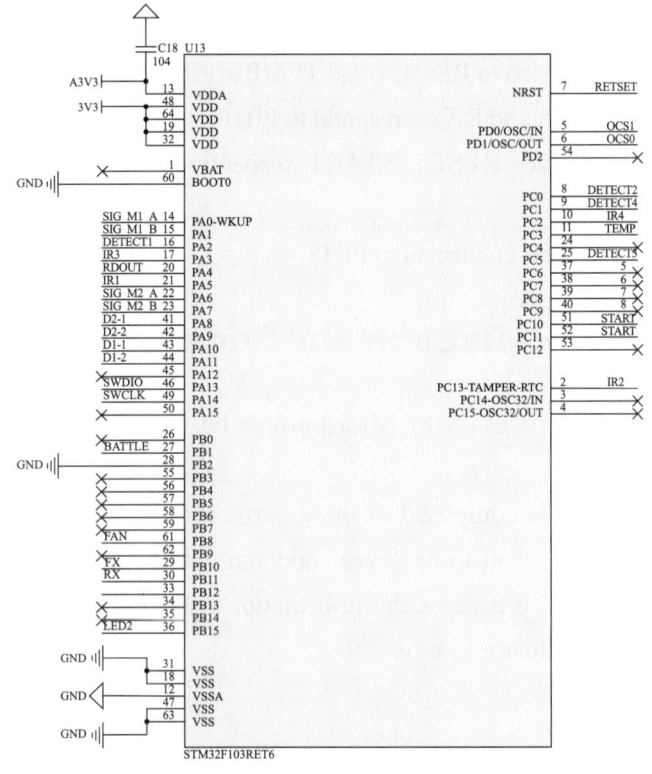

Fig. 1–2–3   TQD-Micromouse-JM Ⅱ minimum system schematic

(1) Motor drive PWM D1_1, D1_2, D2_1, D2_2, respectively corresponding to PA10, PA11, PA8, PA9. D1_1, D1_2, connected to BIN1 and BIN2 of the motor driver chip DRV8848, provide the left motor drive signal. D2_1 and D2_2, connected to AIN1 and AIN2 of the motor driver chip DRV8848, provide the right motor drive signal.

(2) STLINK serial JTAG download ports: SWDIO, SWCLK, corresponding to PA13 and PA14 respectively.

(3) Encoder output signal interface SIG_M1_A, SIG_M1_B, SIG_M2_A, SIG_M2_B, corresponding to PA0, PA1, PA6, PA7, using the internal CPU determines the left and right motor reversing orthogonal code encoder.

(4) Four infrared emission driving signals IR1, IR2, IR3 and IR4, corresponding to PA5, PC13, PA3 and PC2 respectively, provide MOS gate signals, corresponding to the left, the right, the front-left and the front-right in turn;

(5) Four infrared receiving signals DETECT1, DETECT2, DETECT3 and DETECT4, corresponding to PA2, PC0, PC5 and PC1 respectively, are connected to the output of infrared receiving head;

(6) The gyroscope interfaces, RDOUT and TEMP, correspond to PA4 and PC3 respectively, are connected to RDOUT and TEMP of ADXRS620;

(7) Serial ports TX and RX correspond to PB10 and PB11 respectively;

(8) Button interface RESET, START, respectively corresponding to NRST, PC10 and PC11;

(9) LED indicator is connected to PB15.

## Reflection and Summary

(1) Could the processor of Micromouse be replaced by other types of processors?

(2) Micromouse is composed of three parts: sensor, controller and actuator. The IR sensor is equivalent to its "eyes" and can detect the distance of obstacles around. The controller processes the information based on this information, and finally controls the actuator to move.

# Project 3

## Development Environment of Micromouse

 Learning objectives

(1) Understanding IAR EWARM installation and use.

(2) Mastering the method to download program for Micromouse.

# Task 1　IAR EWARM Development Environment

TQD-Micromouse-JM Ⅱ uses IAR Embedded Workbench for ARM (hereinafter referred to as IAR) as the program development environment. It includes project manager, editor, C/C++ compiler and ARM assembler, connector XLINK and RTOS debugging tool C-SPY. In EWARM environment, C/C++ can be used to develop embedded applications conveniently. Compared with other ARM development environments, IAR EWARM has the characteristics of easy entry, convenient use and compact code.

We provide a complete driver library and full labyrinth Demo routines, including the bottom driver, top-level intelligent algorithm, and basic experimental program. Readers can develop as long as they understand the C language.

The software interface is shown in Fig. 1–3–1.

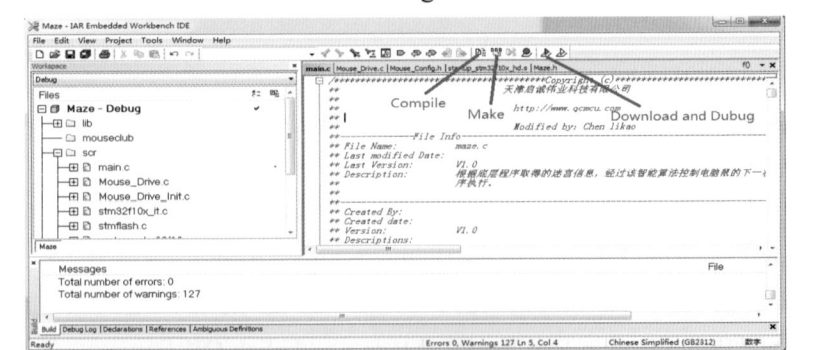

Fig. 1–3–1　IAR software interface

# Task 2  Program Download for Micromouse

## 1. J-Link downloader

J-Link is suitable for debugging and downloading of single-chip programs. This debugger, combined with the IAR EWARM integrated development environment, can support downloading and debugging of all ST series MCU programs.

J-Link uses a USB interface to connect with a computer, whether it is a desktop computer or a notebook computer, it is free to use.

## 2. Connect the hardwares

Be sure to connect Micromouse, downloads and computer correctly before downloading the program. Hardware Connection is shown in Fig. 1–3–2.

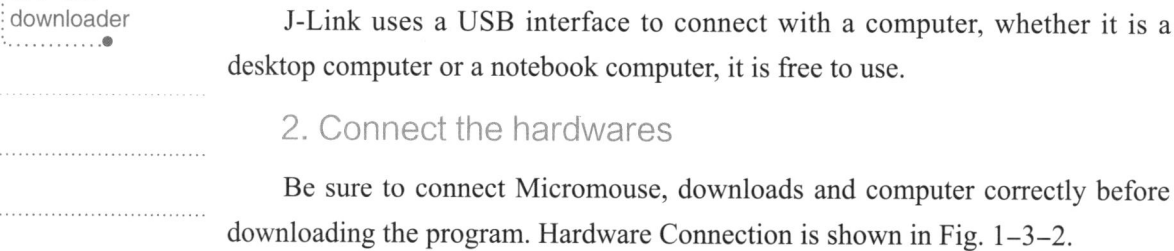

| VCC | 1 • | • 2 | NC |
| Not used | 3 • | • 4 | GND |
| Not used | 5 • | • 6 | GND |
| SWDIO | 7 • | • 8 | GND |
| SWCLK | 9 • | • 10 | GND |
| Not used | 11 • | • 12 | GND |
| SWO | 13 • | • 14 | GND |
| RESET | 15 • | • 16 | GND |
| Not used | 17 • | • 18 | GND |
| 5V-Supply | 19 • | • 20 | GND |

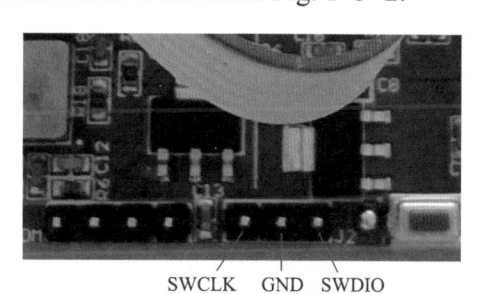

SWCLK    GND   SWDIO

Fig. 1–3–2    Hardware connection

## 3. Download the program

Open the demo, double-click Maze.eww to open the project, as shown in Fig. 1–3–3.

(1) Add TQD library file.

The TQD library file needs to be added before downloading the program, otherwise an error will occur.

Right-click the project and click to enter Options settings. Select the Linker section and enter the Library tab, as shown in Fig. 1–3–4.

Click the red circle icon in Fig. 1–3–4 to add the library file. The library file location is "Demo\Debug\Exe", select Maze.a and click OK button.

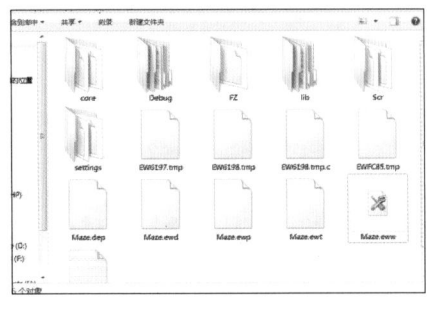 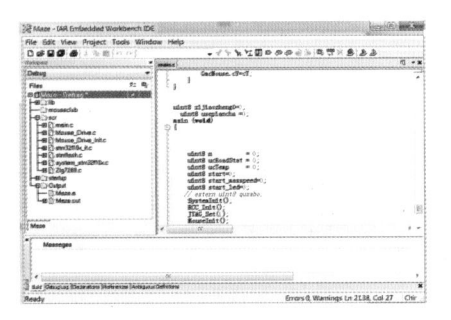

Fig. 1-3-3   Open Demo project

Software •·······

Downloading
the TQD
library

• ··········

Fig. 1-3-4   Add TQD library file

(2) Compile and download. After adding the TQD library file correctly, compile and download again, you will not get an error, as shown is Fig.1-3-5.

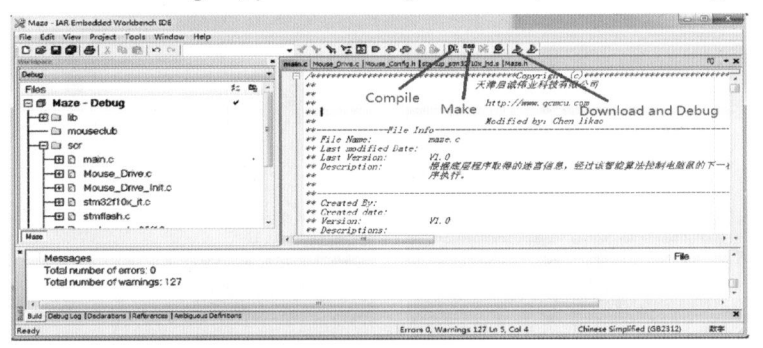

Fig. 1-3-5   Download program

## Reflection and Summary

(1) What other common softwares for developping development C language?

(2) IAR Embedded Workbench provides powerful configuration functions. When downloading the program, you need to select the downloader model and download method according to the actual situation.

# Project 4

## Basic Function Control of Micromouse

### Learning objectives

(1) Learning the infrared detection principle of Micromouse.

(2) Mastering how to drive the Micromouse motor.

(3) Mastering the methods of Micromouse posture detection.

## Task 1 Infrared Detection of Micromouse

Sensors play a very important role in control and are essential components in the sensing system. There are four groups of analog IR sensor on TQD-Micromouse-JM II. Each group of sensors consists of an infrared transmitter and an infrared receiver (see Fig.1–4–1).

● Text

Extended knowledge of sensors

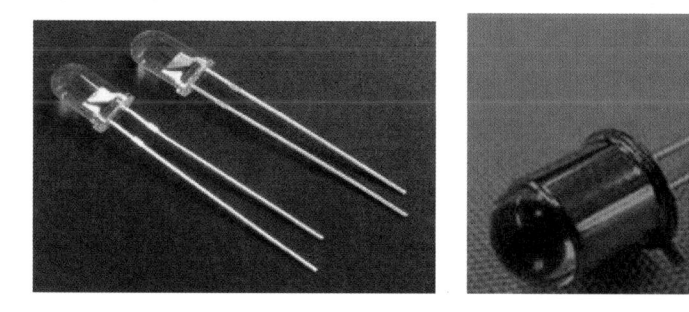

Fig. 1–4–1   TQD-Micromouse-JM II infrared transmitter and receiver

### 1. Micromouse IR sensor transceiver circuit

The Micromouse infrared transceiver sensor circuit is shown in Fig. 1–4–2. IR1-IR4 are infrared emission signals to drive the corresponding infrared emission. The infrared control signal sent by the processor is amplified by the MOS tube to control the on and off of the infrared emission tube to keep the emission frequency of the infrared emission tube at 1 kHz. DETECT1-

DETECT4 are received by the IR sensor. The received infrared reflected signals are respectively connected to the four A/D channels of the processor, and the processor calculates the current distance of Micromouse from the obstacle according to the received voltage value.

The sensor circuit principle is the same in all four directions. The infrared emission head is directly controlled by the GPIO to control whether it is switched on and emitted in order, so as to achieve the purpose of the four groups of sensors not interfering with each other.

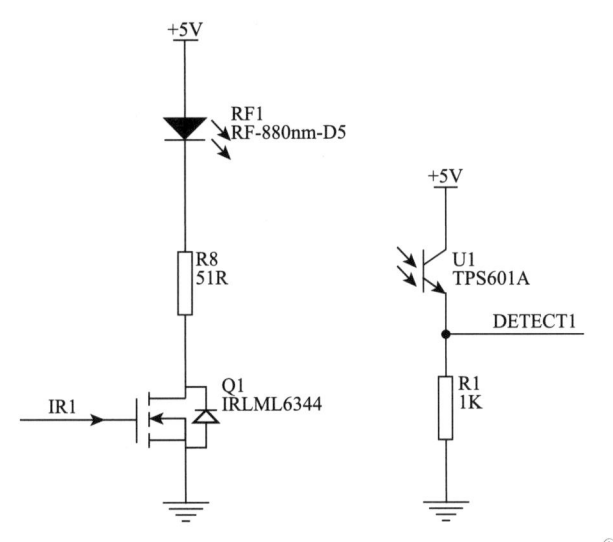

Fig. 1–4–2  TQD-Micromouse-JM II infrared circuit[1]

## 2. IR sensor detection and debugging

Micromouse infrared is used to detect the maze baffle, which is divided into four directions: left front (No. 0), right front (No. 1), left oblique (No. 2), right oblique (No. 3), as shown in Fig. 1–4–3.

Fig. 1–4–3  Infrared grouping

According to the direction of emission, it can be divided into two categories:

Forward Type:

---

① Similar drawings are schematic diagrams derived from Protel 99SE, and their graphic symbols are inconsistent with the national standard symbols. Please refer to the Appendix E for the comparison between the two.

| No. 0 No. 1 | Detect the front, emitting near and far infrared rays of different intensities. Function: 180° turn (near distance), assist 90° turn (far distance). |

Oblique Type:

| No. 2 No. 3 | Detect both sides and emit three kinds of infrared rays with different strengths. Function: Correct vehicle posture at near distance, assist vehicle posture correction at mid distance, and detect intersection at far distance. |

TQD- Micromouse-JM Ⅱ adopts the method of online debugging, in addition to the infrared emission intensity, it also needs to perform the calibration of the Micromouse posture. The calibration of the distance between the baffles on both sides helps Micromouse to correct the vehicle posture deviation when turning.

(1) Distance calibration, as shown in Fig. 1–4–4.

| Step 1: Place Micromouse in the middle of a single channel. Step 2: Download the program and click "Go" to run. Step 3: Calibrate the values of leftdis and rightdis to the values of left_distance and right_distance. See 55 and 71 (integer) as shown in Fig. 1–4–5. |

Fig. 1–4–4　Distance calibration

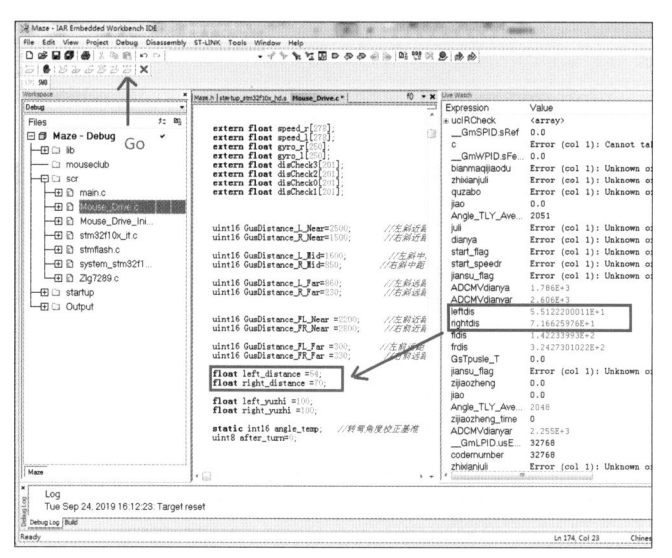

Fig. 1–4–5　Distance calibration data

(2) No. 0 ~ No. 3, four infrared detection intensity calibration:

① Front infrared calibration (No. 0 and No. 1):

Far-distance calibration, as shown in Fig. 1–4–6.

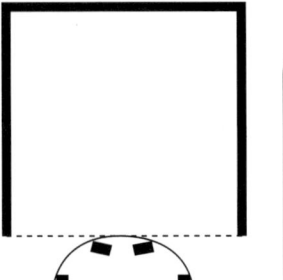

Step 1: Place Micromouse one cell away from the front wall.

Step 2: Keep the program running.

Step 3: Mark the values of No. 0 and No. 1 in the ucIRCheck column as the values of GusDistance_FL_Far and GusDistance_FR_Far, respectively, 753 and 515 as shown in Fig. 1–4–7.

Fig. 1–4–6　Far-distance calibration

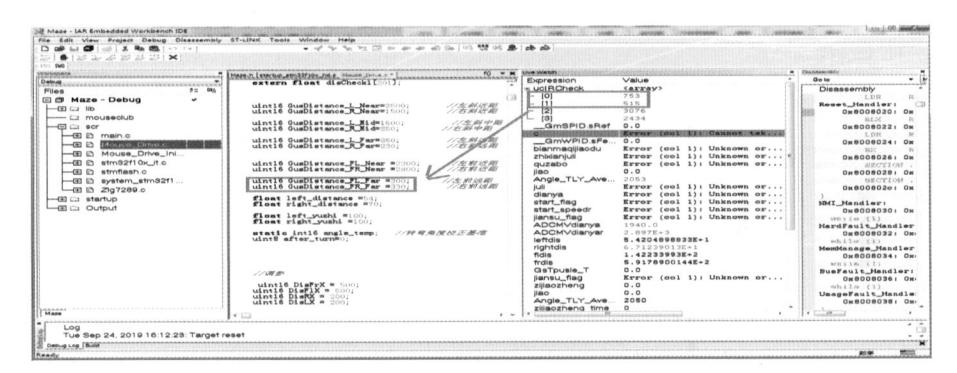

Fig. 1–4–7　Far-distance data ahead

Near-distance calibration, as shown in Fig. 1–4–8.

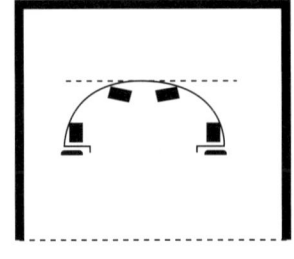

Step 1: Place Micromouse in the middle of a single channel.

Step 2: Keep the program running.

Step 3: Mark the values of No. 2 and No. 3 in the ucIRCheck column as the values of GusDistance_L_Near and GusDistance_R_Near, respectively, 3235 and 2322 as shown in Fig. 1–4–9.

Fig. 1–4–8　Near-distance calibration

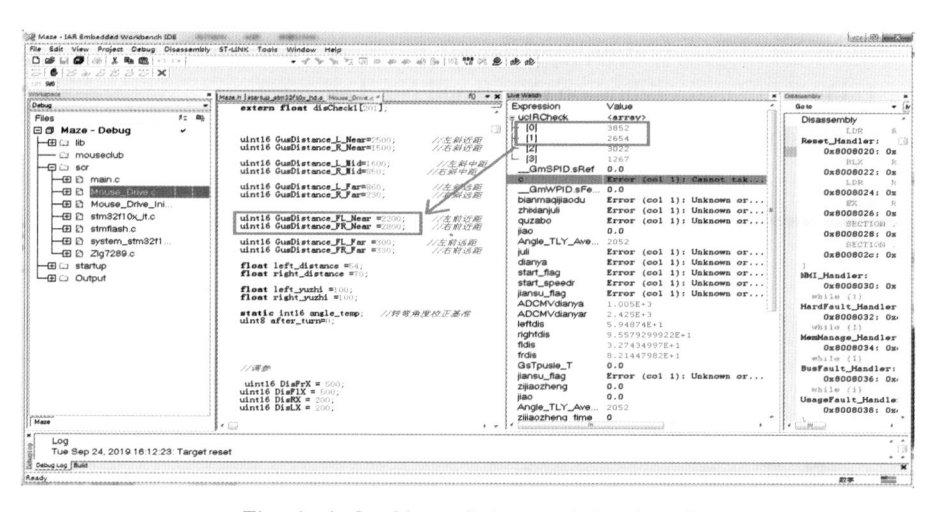

Fig. 1-4-9　Near distance data ahead

② Oblique infrared calibration (No. 2 and No. 3):

Left-oblique and right-oblique near-distance calibration, as shown in Fig.1-4-10.

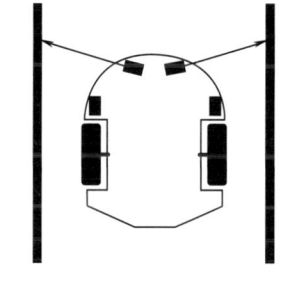

Step 1: Place Micromouse in the middle of a single channel.

Step 2: Keep the program running.

Step 3: Mark the values of No. 2 and No. 3 in the ucIRCheck column as the values of GusDistance_L_Near and GusDistance_R_Near, respectively, 3235 and 2322 as shown in Fig. 1-4-11.

Fig. 1-4-10　Left-oblique and right-oblique near-distance calibration

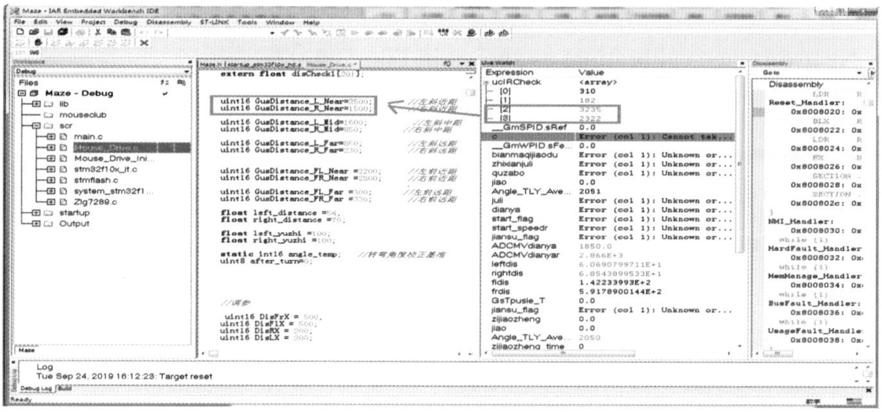

Fig. 1-4-11　Left-oblique and right-oblique near-distance data

Left-oblique and right-oblique far-distance calibration, as shown in Fig.1-4-12.

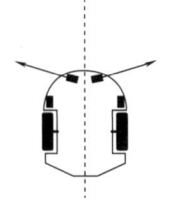

Step 1: Place Micromouse in the middle of two channels.

Step 2: Keep the program running.

Step 3: Mark the values of No. 2 and No. 3 in the ucIRCheck column as the values of GusDistance_L_Fa and GusDistance_R_Far, respectively, 1025 and 411 as shown in Fig. 1–4–13.

Fig. 1–4–12　Left-oblique and right-oblique far-distance calibration

Video

Micromouse Infrared Detection

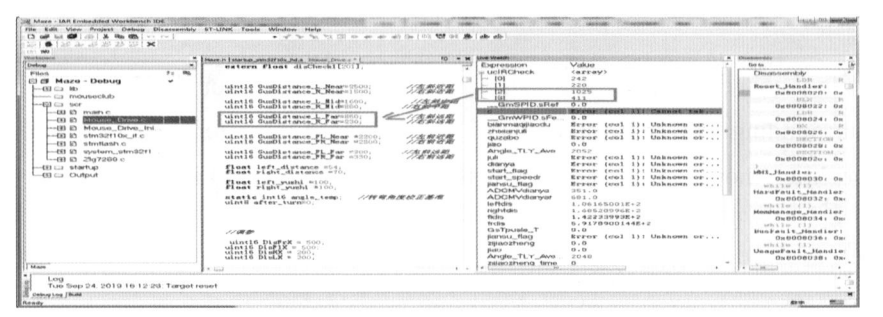

Fig. 1–4–13　Left-oblique and right-oblique far-distance data

Middle distance: Divide the sum of near distance and far distance (GusDistance_L_Mid and GusDistance_R_Mid) by 2 as the middle distance.

## Experiment　Micromouse infrared detection

Core function 1: __irSendFreq

```
/******************************************************************
** Function name:__irSendFreq
** Descriptions:main function
** input parameters:None
** output parameters:None
** Returned value:None
******************************************************************/
void __irSendFreq (int8 __cNumber)
{
    switch (__cNumber)
    {
      case 1:
          GPIO_SetBits(GPIOA,GPIO_Pin_5);//Drive the left infrared detection
           break;
      case 2:
          GPIO_SetBits(GPIOC,GPIO_Pin_13);//Drive the right infrared detection
           break;
      case 3:
```

```
                    GPIO_SetBits(GPIOA,GPIO_Pin_3); //Drive the front-left
infrared detection
                break;
            case 4:
                GPIO_SetBits(GPIOC,GPIO_Pin_2); // Drive the front-right
infrared detection
                break;
            default:
                break;
        }
    }
```

Flow chart. In this experiment, four groups of IR sensor are driven to detect the surrounding wall information, and the change of the return value is observed online. Micromouse infrared detection flow chart is shown in Fig. 1–4–14.

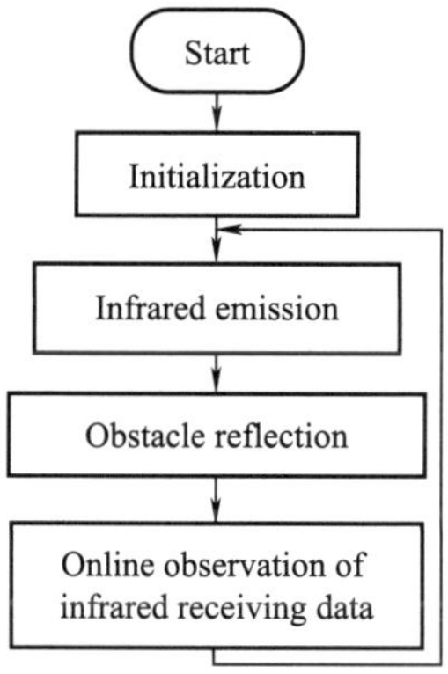

Fig. 1–4–14   Micromouse infrared detection flow chart

Main program:

main.c

```
/*************************************************************
** Function name:main
** Descriptions:The main program
** input parameters:None
** output parameters:None
** Returned value:None
*************************************************************/
main (void)
{
    uint8 n=0; /*The number of coordinates with multiple branches*/
    uint8 ucRoadStat=0; /*Counting the number of directions can
move forward*/
    uint8 ucTemp=0; /*Used for coordinate conversion in START state*/
```

```
uint8 start=0;
uint8 start_maxspeed=0;
uint8 start_led=0;
SystemInit();
RCC_Init();
JTAG_Set(1);
MouseInit();
    ⋮

        while (1)
        {
            if (startCheck()==true)
            {
                break;
            }
        }
        break;
    default:
        break;
    }
}
}
```

After downloading the program, observe the return value of the four groups of IR sensor online.

# Task 2   Motor Drive of Micromouse

## 1. DC motor

TQD-Micromouse-JM II DC motor uses FAULHABER1717SR coreless DC motor (see Fig. 1–4–15), input voltage is 7.4 V, can output 1.96 W power, maximum speed is 14,000 r/min, its main parameters are shown in Table 1–4–1, the dimensions as shown in Fig. 1–4–16.

Text

Extended knowledge of motors

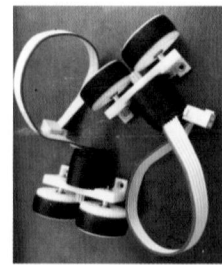

Fig. 1–4–15    FAULHABER1717SR coreless DC motor

## Table 1–4–1　The main parameters of coreless DC motor

| Series 1717 ... SR | | 1717 T | 003 SR | 006 SR | 012 SR | 018 SR | 024 SR | |
|---|---|---|---|---|---|---|---|---|
| 1 Nominal voltage | $U_N$ | | 3 | 6 | 12 | 18 | 24 | V |
| 2 Terminal resistance | R | | 1,07 | 4,3 | 17,1 | 50,1 | 68,8 | Ω |
| 3 Output power | $P_{2\ max.}$ | | 1,97 | 1,96 | 1,97 | 1,5 | 1,96 | W |
| 4 Efficiency, max. | $\eta_{max}$ | | 69 | 69 | 70 | 68 | 70 | % |
| | | | | | | | | |
| 5 No-load speed | $n_0$ | | 14 000 | 14 000 | 14 000 | 12 300 | 14 000 | rpm |
| 6 No-load current (with shaft ø 1,5 mm) | $I_0$ | | 0,091 | 0,046 | 0,023 | 0,013 | 0,011 | A |
| 7 Stall torque | $M_H$ | | 5,37 | 5,34 | 5,38 | 4,66 | 5,36 | mNm |
| 8 Friction torque | $M_R$ | | 0,18 | 0,18 | 0,18 | 0,18 | 0,17 | mNm |
| | | | | | | | | |
| 9 Speed constant | $k_n$ | | 4 820 | 2 410 | 1 210 | 709 | 602 | rpm/V |
| 10 Back-EMF constant | $k_E$ | | 0,207 | 0,414 | 0,829 | 1,41 | 1,66 | mV/rpm |
| 11 Torque constant | $k_M$ | | 1,98 | 3,96 | 7,92 | 13,5 | 15,9 | mNm/A |
| 12 Current constant | $k_I$ | | 0,505 | 0,253 | 0,126 | 0,074 | 0,063 | A/mNm |
| | | | | | | | | |
| 13 Slope of n-M curve | $\Delta n/\Delta M$ | | 2 610 | 2 620 | 2 600 | 2 640 | 2 610 | rpm/mNm |
| 14 Rotor inductance | L | | 17 | 65 | 260 | 760 | 1 040 | µH |
| 15 Mechanical time constant | $\tau_m$ | | 16 | 16 | 16 | 16 | 16 | ms |
| 16 Rotor inertia | J | | 0,59 | 0,58 | 0,59 | 0,58 | 0,59 | gcm² |
| 17 Angular acceleration | $\alpha_{max.}$ | | 92 | 92 | 92 | 80 | 92 | ·10³rad/s² |
| | | | | | | | | |
| 18 Thermal resistance | $R_{th1}$ / $R_{th2}$ | 4,5 / 27 | | | | | | K/W |
| 19 Thermal time constant | $\tau_{w1}$ / $\tau_{w2}$ | 2 / 210 | | | | | | s |
| 20 Operating temperature range: | | | | | | | | |
| – motor | | -30 ... +85 (optional version   -55 ... +125) | | | | | | °C |
| – rotor, max. permissible | | +125 | | | | | | °C |
| | | | | | | | | |
| 21 Shaft bearings | | sintered bearings | | ball bearings | | ball bearings, preloaded | | |
| 22 Shaft load max.: | | (standard) | | (optional version) | | (optional version) | | |
| – with shaft diameter | | 1,5 | | 1,5 | | 1,5 | | mm |
| – radial at 3 000 rpm (3 mm from bearing) | | 1,2 | | 5 | | 5 | | N |
| – axial at 3 000 rpm | | 0,2 | | 0,5 | | 0,5 | | N |
| – axial at standstill | | 20 | | 10 | | 10 | | N |
| 23 Shaft play | | | | | | | | |
| – radial | ≤ | 0,03 | | 0,015 | | 0,015 | | mm |
| – axial | ≤ | 0,2 | | 0,2 | | 0 | | mm |
| | | | | | | | | |
| 24 Housing material | | steel, black coated | | | | | | |
| 25 Weight | | 18 | | | | | | g |
| 26 Direction of rotation | | clockwise, viewed from the front face | | | | | | |
| **Recommended values** - mathematically independent of each other | | | | | | | | |
| 27 Speed up to | $n_{e\ max.}$ | | 10 000 | 10 000 | 10 000 | 10 000 | 10 000 | rpm |
| 28 Torque up to | $M_{e\ max.}$ | | 2 | 2 | 2 | 2 | 2 | mNm |
| 29 Current up to (thermal limits) | $I_{e\ max.}$ | | 1,2 | 0,6 | 0,3 | 0,18 | 0,15 | A |

Orientation with respect to motor terminals not defined

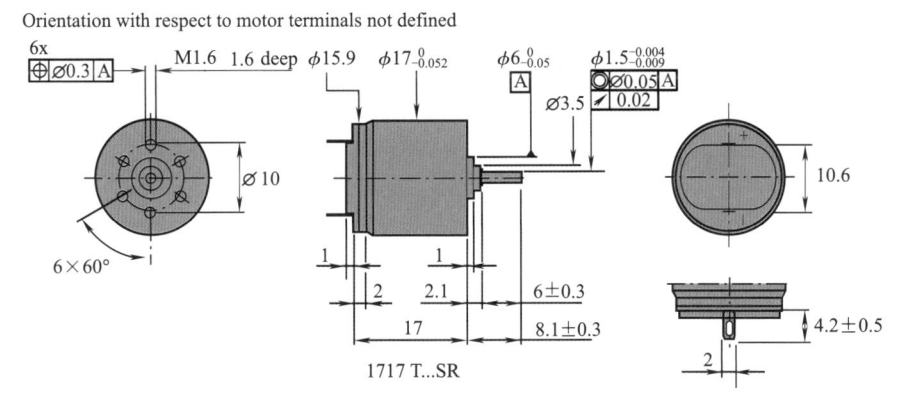

1717 T...SR

Fig. 1–4–16　Coreless DC motor dimensions

## 2. Driver chip

The motor drive uses the DRV8848 chip, and the circuit diagram is shown in Fig. 1–4–17.

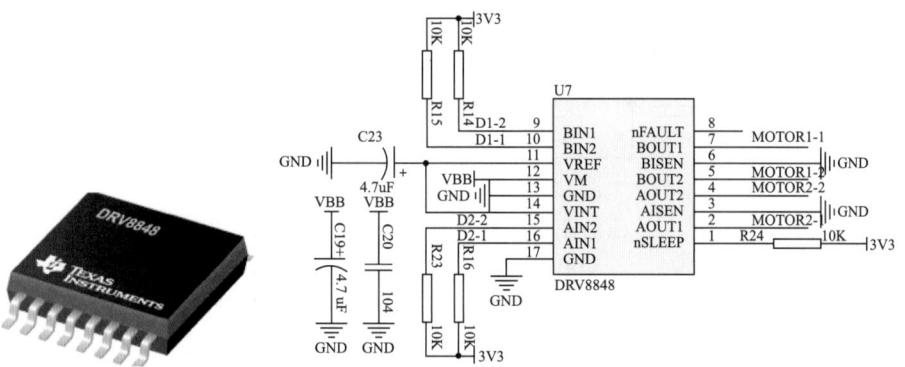

Fig. 1–4–17　Motor drive principle diagram

The DRV8848 contains a total of two H-bridge drivers, each of them contains N-channel and P-channel power MOSFET configured as full H-bridges to drive the motor windings. Each H-bridge contains an adjustment circuit to adjust the winding current through a fixed off-time chopping scheme. DRV8848 can output up to 2A current from each H-bridge and drive up to 4 A current in parallel mode (normal heat dissipation, 12 V and TA=25 ℃). It can achieve good motor start, brake, forward and reverse control. DRV8848 truth table is shown in Table 1–4–2.

Table 1–4–2　DRV8848 truth table

| PWM_H | PWM_L | State |
|-------|-------|-------|
| L | L | Forward brake |
| L | H | Forward |
| H | L | Reverse |
| H | H | Reverse brake |

Four H-bridge drive modes are shown in Fig.1–4–18.

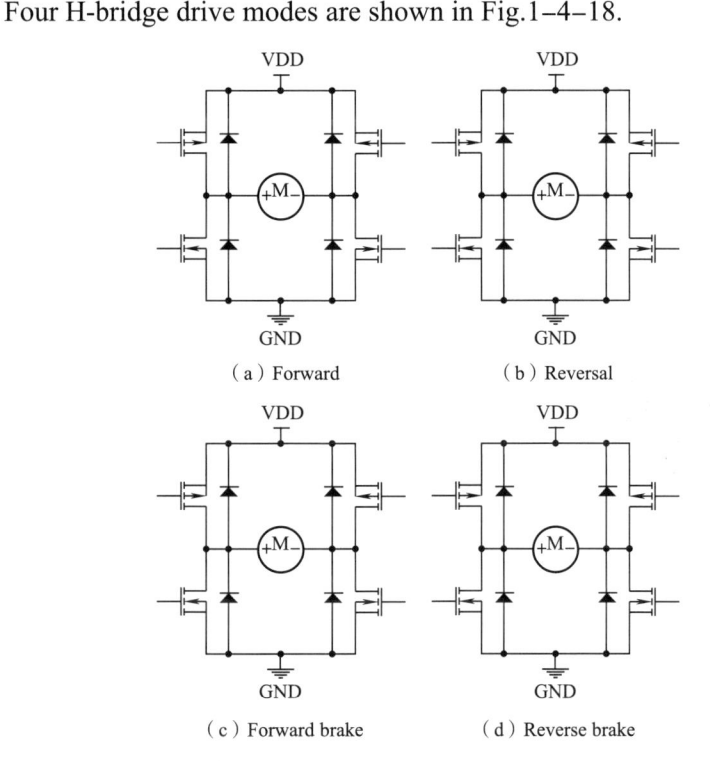

( a ) Forward          ( b ) Reversal

( c ) Forward brake          ( d ) Reverse brake

Fig. 1–4–18　Four H-bridge drive modes

## 3. PWM drive speed regulation

The Fig. 1–4–19 is the most basic PWM (pulse width modulation) driving motor circuit. When the gate ($U_i$ input terminal) of the switching tube V1 inputs a high level, V1 is turned on, and the voltage across the armature winding of the motor is $U_s$. After $t_1$, the gate input becomes low level, V1 is cut off, the self-inductance current of the motor is quickly released through the freewheeling diode D1, and the voltage across the armature becomes 0. After $t_2$, the gate input back to high level again, and V1 repeats the previous process. The average voltage $U_0$ across the armature winding of the motor is obtained as shown in the following formula.

$$U_0=(t_1 \cdot U_s)/(t_1+t_2)=U_s \cdot (t_1/T)=\alpha \cdot U_s$$

$\alpha$ in the above formula is the duty cycle.

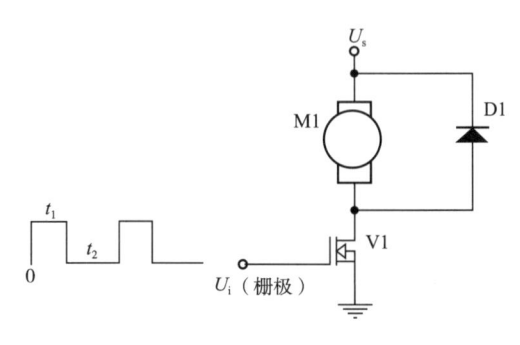

Fig. 1-4-19  Basic PWM drive

Changing the armature voltage is the main method of DC speed regulation. TQD-Micromouse-JM II adopts PWM speed regulation method, and changes the applied average voltage $U_d$ by adjusting the duty cycle of the PWM trigger signal of the microprocessor, thus achieving the speed regulation of the DC motor. Use unipolar PWM signal to drive DC motor. It controls the forward and reverse PWM waveforms of the motor as shown in Fig. 1-4-20, and adjusts the speed of the motor by changing the PWM duty cycle.

Fig. 1-4-20  PWM wave forms

Video

Experiment:
Micromouse
Runs

## Experiment  Micromouse runs

Through the above knowledge learning, we knew that the motor has many parameters to consider:

(1) Motor status: start or stop?

(2) Running direction: forward or backward?

(3) Speed: fast or slow?

(4) The number of steps need to rotate.

(5) The number of steps that have been turned.

Thus, we can build a function structure to store these parameters.

The structure is the same as other basic data types (such as int type, char

type), but the structure can be customized as needed. The role of the structure in the function is encapsulation. The advantage of encapsulation is that it can be used again. Users do not have to care about what the structure is, as long as it is used according to the definition.

The Structure definition: Motor drive.

```
/************************************************************
Acceleration and deceleration
************************************************************/
#define__SPEEDUP           0           /*Acceleration*/
#define__SPEEDDOWN         1           /*Deceleration*/
/************************************************************
Motor State
************************************************************/
#define__MOTORSTOP         0           /*Motor stop*/
#define__WAITONESTEP       1           /*Wait one step*/
#define__MOTORRUN          2           /*Motor run*/
/************************************************************
Motor direction
************************************************************/
#define__MOTORGOAHEAD      0           /*Motor go ahead*/
#define__MOTORGOBACK       1           /*Motor go back*/
#define__MOTORGOSTOP       2           /*Motor stop*/
/************************************************************
Structure definition
************************************************************/
struct__motor {
    int8    cState;         /*Motor state*/
    int8    cDir;           /*Motor running direction*/
    int8    cRealDir;       /*Absolute direction*/
    uint32  uiPulse;        /*Pulses required for motor running*/
    uint32  uiPulseCtr;
                    /*Pulses have generated by motor running*/
    int16   sSpeed;         /*Current duty cycle*/
};
typedef struct__motor__MOTOR;
            Core function 1:__rightMotorContr

/************************************************************
** Function name:__rightMotorContr
** Descriptions:Right DC motor drive
** input parameters:_GmRight.cDir(Motor running direction)
** output parameters:None
** Returned value:None
************************************************************/
void__rightMotorContr(int speed)
```

```
{
    switch (__GmRight.cDir)
    {
    case__MOTORGOAHEAD:
      TIM_SetCompare1(TIM1,speed);
      TIM_SetCompare2(TIM1,0);
        break;

    case__MOTORGOBACK:
      TIM_SetCompare1(TIM1,0);
      TIM_SetCompare2(TIM1,speed);
        break;

    case __MOTORGOSTOP:
      TIM_SetCompare1(TIM1,0);
      TIM_SetCompare2(TIM1,0);
        break;

    default:
        break;
    }
}
               Core function 2:__leftMotorContr
/****************************************************************
** Function name:__leftMotorContr
** Descriptions:Left DC motor drive
** input parameters:_GmLeft.cDir(Motor running direction)
** output parameters:None
** Returned value:None
****************************************************************/
void__leftMotorContr(int speed)
{
    switch (__GmLeft.cDir)
    {
    case__MOTORGOAHEAD:
      TIM_SetCompare4(TIM1,0);
      TIM_SetCompare3(TIM1,speed);
        break;

    case__MOTORGOBACK:
      TIM_SetCompare4(TIM1,speed);
      TIM_SetCompare3(TIM1,0);
        break;

    case__MOTORGOSTOP:
      TIM_SetCompare3(TIM1,0);
```

```
        TIM_SetCompare4(TIM1,0);
          break;

    default:
        break;
    }
}
```

Flow chart. This experiment learns the driving of the DC motor and controls the operation of Micromouse at the same speed and differential speed through the different number of buttons. The flow chart is shown in Fig. 1–4–21.

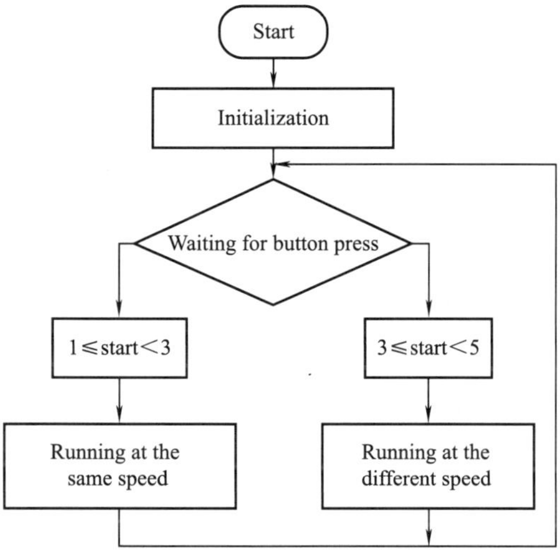

Fig. 1–4–21   Speed control flow chart

Main program:

```
                            main.c

/*************************************************************
** Function name:main
** Descriptions:Motor drive
** input parameters:None
** output parameters:None
** Returned value:None
*************************************************************/
main (void)
{
    uint8  n=0;   /*The number of coordinates with multiple branches*/
    uint8  ucRoadStat=0;
        /*Counting the number of directions can move forward*/
    uint8  ucTemp=0;  /*Used for coordinate conversion in START state*/
```

```
uint8 start=0;
uint8 start_maxspeed=0;
uint8 start_led=0;
SystemInit();
RCC_Init();
JTAG_Set(1);
MouseInit();
PIDInit();
ZLG7289Init();
delay(100000);
delay(100000);
GPIO_Config1();
USART1_Config1();
NVIC_Config1();
floodwall();
GPIO_SetBits(GPIOB,GPIO_Pin_12);
while (1) {
  if (startCheck() == true)
  {
    start=1;
  }
  if((start==1)&&GucGoHead)
  {
    __rightMotorContr(200);
    __leftMotorContr(200);
  }
  if((start==1)&&GucGoHead1)
  {
    __rightMotorContr(400);
    __leftMotorContr(0);
  }
 }
}
```

# Task 3   Attitude Detection of Micromouse

Because the speed of the coreless DC motor is extremely fast, the turning angle of Micromouse cannot achieve the purpose of precise control simply relying on the time or pulse number of differential operations. Therefore, in this task we learn to use the gyroscope to control the turning angle of Micromouse.

TQD-Micromouse-JM II uses ADXRS620 gyroscope to detect the rotation angle as shown in Fig. 1–4–22. ADXRS620 is a linear Z-axis angular velocity

sensor whose output voltage is proportional to the angular velocity. By integrating the angular velocity, the angle value can be obtained. There by precisely controlling the turning angle.

● Video

Experiment:
attitude
detection

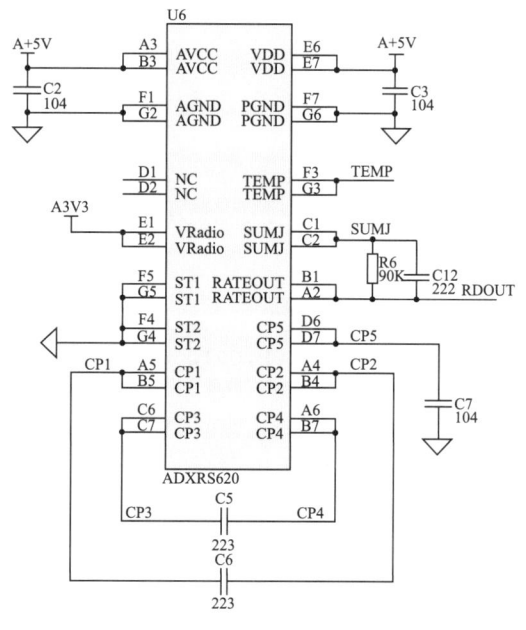

Fig. 1-4-22　ADXRS620 physical diagram and schematic diagram

The angular velocity reference axis of the gyroscope ADXRS620 is shown in Fig. 1-4-23. The output voltage of the RATEOUT is proportional to the angular velocity. When rotating clockwise, the angular velocity of the reference axis is a positive value, otherwise it is a negative value.

Fig. 1-4-23　Measurement diagram of ADXRS620

The reference voltage of the gyroscope ADXRS620 is 2.5 V; the relationship between the output voltage and the angular velocity is 6mV/(°); when the output voltage is greater than the reference voltage (2.5 V), the angular velocity is

positive (clockwise), otherwise it is negative. Assuming that $v$ is the rotation speed of the gyroscope, and the unit is $(°)/s$, then the relationship between the output voltage of the gyroscope and the rotation speed $v$ is:

$$RATEOUT=2.5+0.006 \text{ V}$$

Assuming $v=300°/s$, then

$$RATEOUT=(2.5+0.006\times300) \text{ V}=4.3V$$

Assuming $v=-300°/s$, then

$$RATEOUT=(2.5-0.006\times300) \text{ V}= 0.7 \text{ V}$$

Calculation of angular velocity:

$$v=1000(RATEOUT-2.5)/6$$

$$\frac{v}{\omega}=\frac{360}{2\pi} \rightarrow \omega=2\pi v/360 \text{ rad/s}$$

Angle calculation:

$$\theta = \int_0^t \omega dt = \int_0^t 2\pi v/360 dt$$

Calculation method related to programming: We use a ten-bit ADJI to divide the 0-5 V gyro voltage into 1,023 parts ($2^{10}=1,024$, and then subtract 1). Thus, the digital voltage of A/D corresponding to 2.5 V is 511, then:

$$\omega = 2\pi\times1,000(RATEOUT-511)/6/360$$

$$\omega=\frac{25\times\pi}{27}\times(RATEOUT - 511)$$

$$\theta = \int_0^t wdt = \sum w(i)(t)i$$

```
                Core function 1: voltageDetect
/******************************************************************
** Function name:voltageDetect
** Descriptions:Gyro voltage collection
** input parameters:None
** output parameters:None
** Returned value:None
******************************************************************/
void voltageDetect(void)
{
  u16 w;
  if(Angle_TLY_Average>=voltageDetectRef)
  {
```

```
        w=Angle_TLY_Average-voltageDetectRef;
    }
    else
    {
        w= voltageDetectRef - Angle_TLY_Average;
    }
    GW=GW+w;
}
```

Flow chart. In this experiment, we can observe the changes of gyroscope parameters online by artificially rotating Micromouse. The flow chart is shown as Fig. 1-4-24.

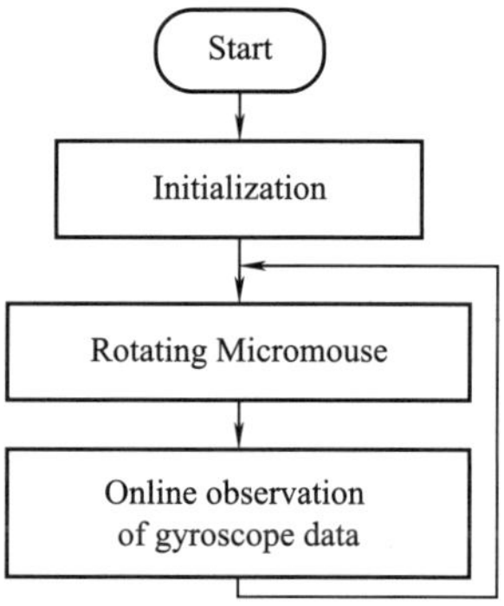

Fig. 1-4-24    Observe the output value of the gyroscope flow chart

Main program:

main.c

```
/*************************************************************
** Function name:main
** Descriptions:Collection and analysis of gyro voltage
** input parameters:None
** output parameters:None
** Returned value:None
*************************************************************/
main (void)
{
    uint8 n=0;
            /*The number of coordinates with multiple branches*/
    uint8 ucRoadStat=0;
     /*Counting the number of directions can move forward*/
    uint8 ucTemp=0;
                /*Used for coordinate conversion in START state*/
    uint8 start=0;
```

```
      uint8 start_maxspeed=0;
      uint8 start_led=0;
      SystemInit();
      RCC_Init();
      JTAG_Set(1);
      MouseInit();
      .
      .
      .
            while (1)
            {
                if(startCheck()==true)
                {
                    break;
                }
            }
            break;
        default:
            break;
        }
    }
}
```

Observe the return value of the gyroscope online after downloading the program.

# Chapter 2

# Comprehensive Practice

This chapter mainly introduces the debugging methods of straight movement control and accurate turning control of Micromouse, and explains the advanced functions of Micromouse.

# Project 1

## Motion Attitude Control of Micromouse

 Learning objectives

(1) Mastering the principle and control method of Micromouse speed regulation.

(2) Mastering the method of Micromouse accurate turning control.

The operation of Micromouse in a maze can be simplified into two parts: straight movement and turning. Straight movement refers to Micromouse in the maze through the detection of two side walls to correct it's attitude and avoid touching. Turning refers to the Micromouse precise turning of 180° and 90°, combined with straight movement to the final destination.

## Task 1   Closed-Loop Control of Micromouse Speed

In order to complete the maze search task, Micromouse must have a stable straight speed. TQD-Micromouse-JM II using PWM to adjust the speed of the motor, using an encoder to record the number of motor pulses per unit time to measure the speed of the motor.

However, due to the instability of the system, such as the attenuation of battery voltage, the drift of ambient temperature, the change of device performance and other factors, Micromouse cannot obtain stable speed. To solve this problem, a negative feedback closed-loop control algorithm can be used to control the motor speed.

PID control algorithm (also known as PID controller) is a common motor speed control algorithm. As the earliest practical controller, PID controller has a history of more than 70 years and it is still the most widely used industrial controller. PID controller is simple and easy to understand, and does not need to

use the precise system model and other preconditions, so it has become the most widely used controller.

## 1. PID control principle

PID is the abbreviation of proportion, integral and differential, which respectively represent three control algorithms. The combination of these three algorithms can effectively correct the deviation of the controlled object, so as to achieve a stable state.

The principle of conventional PID control system is shown in Fig. 2–1–1. $t$ represents a certain moment, $r(t)$ is the given value, and $y(t)$ is the actual output value of the system. The given value and the actual output value constitute the control deviation $e(t)$, so:

$$e(t)=r(t)-y(t) \tag{2-1-1}$$

$e(t)$ is used as the input of PID control, $u(t)$ is used as the output of PID controller and the input of controlled object. Therefore, the control law of analog PID controller is as follows:

$$u(t)=K_{p}\left[e(t)+\frac{1}{T_{i}}\int_{0}^{t}e(t)\mathrm{d}t+T_{d}\frac{\mathrm{d}e(t)}{\mathrm{d}t}\right] \tag{2-1-2}$$

Among them, $K_{p}$ is the proportional coefficient; $T_{i}$ is the integral time, also known as the integral coefficient; $T_{d}$ is the differential time, also known as the differential coefficient.

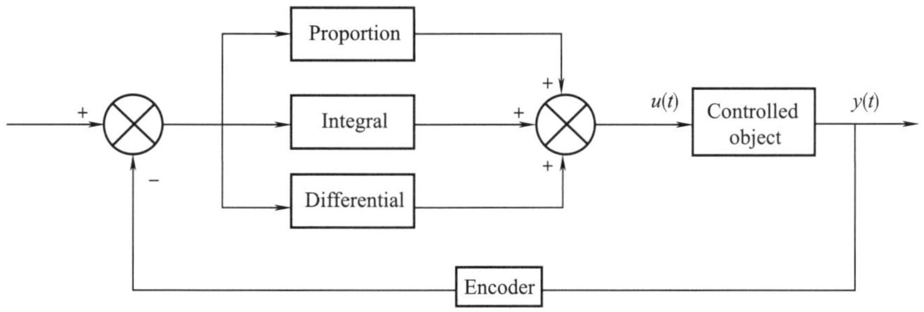

Fig. 2–1–1　Schematic diagram of PID control system

### 1) Proportion part (P)

The proportion part is mainly to quickly control the deviation signal of the control system through proportional response to reduce the change of the deviation signal. $K_{p}e(t)$is the mathematical formula of proportion part, which

needs reasonable adjustment the value of $K_p$. If $K_p$ is too large, the more unstable the system will be. If $K_p$ is too small, the adjustment speed of the system will be slower. The disadvantage of proportional regulation is that it will produce static deviation, that is, when the steady state is reached, the controlled will produce a fixed fluctuation, as shown in Fig. 2–1–2.

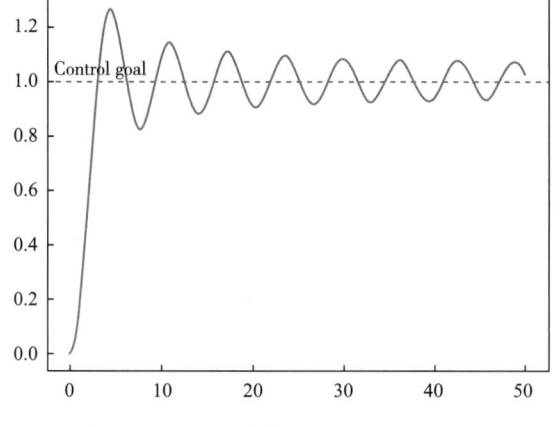

Fig. 2–1–2   Effect of static error

2) Integral part (I)

The integral part can eliminate static deviation, $K_p \dfrac{1}{T_i} \int_0^t e(t)\mathrm{d}t$ is the mathematical expression of integral part. As you can see from the expression, as long as there is a deviation, the integral will play a controlling role until the deviation is zero, as shown in Fig. 2–1–3. The time constant $T_i$ plays an important role in the integral part. When the value of $T_i$ increases, the integral plays a smaller role, and the system deviation removal time becomes longer. When the $T_i$ value decreases, the integral plays a greater role and the system deviation removal time is shortened.

3) Differential part (d)

The function of differential part is to give appropriate correction in advance according to the variation trend of deviation. The introduction of differential action will help to reduce overshoot, overcome oscillation and stabilize the system. It accelerates the tracking speed of the system. The role of differential part is shown in Fig. 2–1–4. But the function of differential is very sensitive to the noise of input signal, and it is not used in the system with large noise.

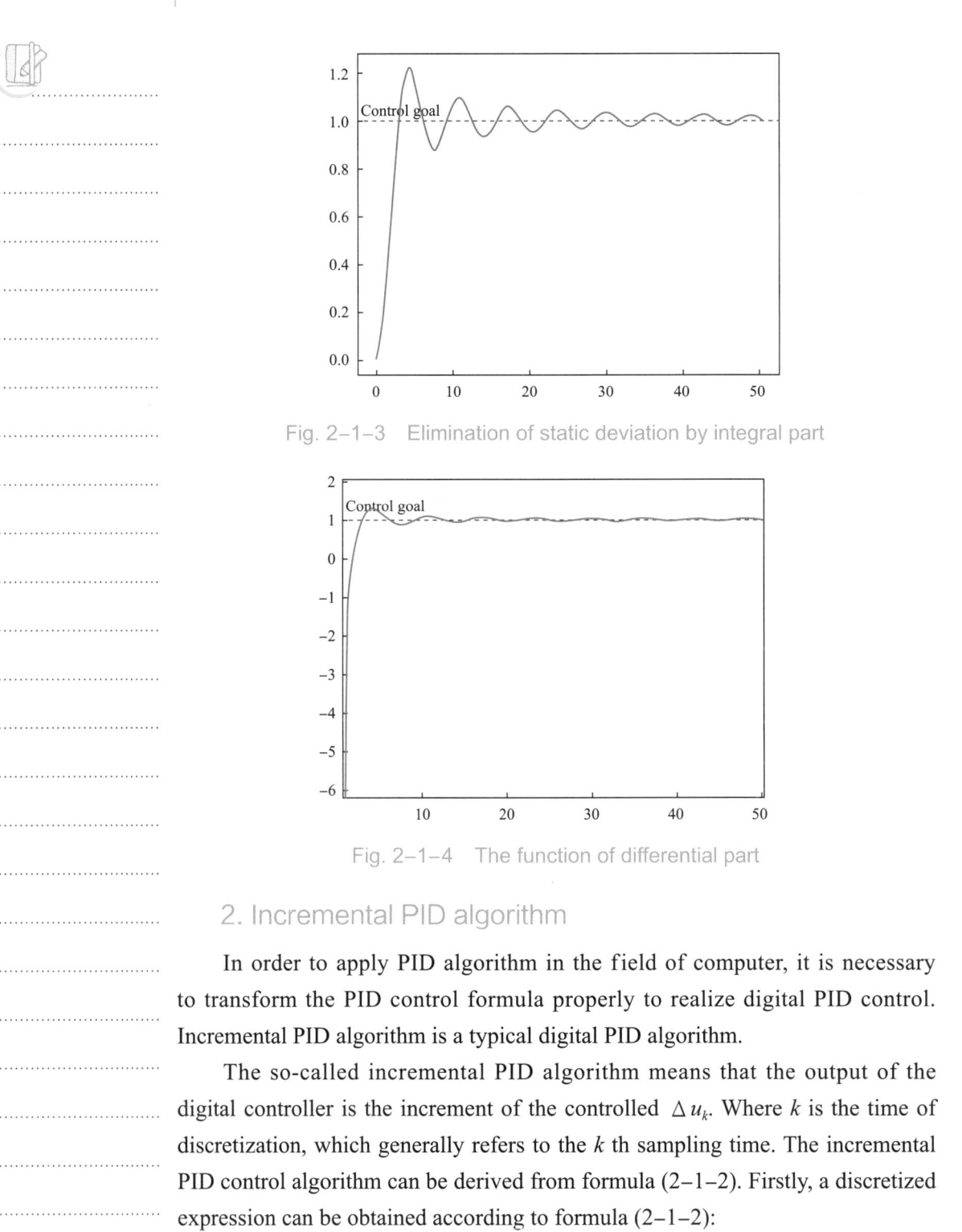

Fig. 2–1–3　Elimination of static deviation by integral part

Fig. 2–1–4　The function of differential part

## 2. Incremental PID algorithm

In order to apply PID algorithm in the field of computer, it is necessary to transform the PID control formula properly to realize digital PID control. Incremental PID algorithm is a typical digital PID algorithm.

The so-called incremental PID algorithm means that the output of the digital controller is the increment of the controlled $\Delta u_k$. Where $k$ is the time of discretization, which generally refers to the $k$ th sampling time. The incremental PID control algorithm can be derived from formula (2–1–2). Firstly, a discretized expression can be obtained according to formula (2–1–2):

$$u_k=K_p\left[e_k+\frac{1}{T_i}\sum_{j=0}^{k}e_j+T_d\frac{e_k-e_{k-1}(t)}{T}\right] \qquad (2-1-3)$$

Among them, $j$ is the cyclic variable used to calculate the integral value from time 0 to $k$; $u_k$ is the output value of $k$ sampling time, $e_k$ is the sampling time error of $k$ times; $u_{k-1}$ is $k-1$ sampling time error; $T$ is sampling period.

According to formula (2–1–3), the difference between the output at time $k$ and that at time $k$-1 can be calculated $\Delta u_k$, as shown in formula (2–1–4):

$$\Delta u_k=u_k-u_{k-1}=K_p\left[e_k-e_{k-1}+\frac{1}{T_i}e_k+T_d\frac{e_k-2e_{k-1}+e_{k-2}}{T}\right]$$

$$=K_p\left(1+\frac{T}{T_i}+\frac{T_d}{T}\right)e_k-K_p\left(1+\frac{2T_d}{T}\right)e_{k-1}+K_p\frac{T_d}{T}e_{k-2}$$

$$=Ae_k-Be_{k-1}+Ce_{k-2} \qquad (2-1-4)$$

Among them, $A=K_p\left(1+\frac{T}{T_i}+\frac{T_d}{T}\right)$; $B=K_p\left(1+\frac{2T_d}{T}\right)$; $C=K_p\frac{T_d}{T}$.

It can be seen from formula (2–1–4) that if the constant period $T$ is adopted, once $A$, $B$ and $C$ are determined, the controlled can be obtained as long as the deviation values of the three measurements before and after are used, and the calculation is relatively small. Finally, we can get the following results:

$$u_k=\Delta u_k+u_{k-1} \qquad (2-1-5)$$

## Experiment   Realization of PID algorithm

Macro definition and global variables.

```
#define__KP 30      //Proportion
#define__KI 0.1     //Integral
#define__KD 0       //Differential
struct__pid          //Define the structure type for PID algorithm
{
    int16 usFeedBack;       //Speed feedback
    uint16 usEncoder_new;   //Encoder
    uint16 usEncoder_last;  //Encoder
    float sRef;
    float sFeedBack;
    float sPreError;   // Speed error, vi_Ref - vi_FeedBack
    float sPreDerror;  // Error of speed error, d_error-PreDerror
    fp32 fKp;          // p
    fp32 fKi;          // I
    fp32 fKd;          // D
    int16 iPreU;       // Motor control value
};
```

Video

Experiment: realization of PID algorithm

```
typedef struct__pid__PID;

__PID__GmSPID;//Declare the PID variable used to control the linear speed
__PID__GmWPID;//Declare the PID variable used to control the turning speed
                    Core function 1: PIDInit
/****************************************************************
** Function name:PIDInit
** Descriptions:PID initialization
** input parameters:None
** output parameters:None
** Returned value:None
****************************************************************/
void PIDInit(void)
{
    __GmLPID.usEncoder_new=32768;
    __GmLPID.usFeedBack=0;  // Speed feedback
    __GmLPID.sFeedBack=0;
    __GmRPID.usEncoder_new=32768;
    __GmRPID.usFeedBack=0;  // Speed feedback
    __GmRPID.sFeedBack=0;
    __GmSPID.sRef=0;          // Speed setting
    __GmSPID.sFeedBack=0;
    __GmSPID.sPreError=0;
    __GmSPID.sPreDerror=0 ;
    __GmSPID.fKp=__KP;
    __GmSPID.fKi=__KI;
    __GmSPID.fKd=__KD;
    __GmSPID.iPreU=0;
    __GmWPID.sRef=0;
    __GmWPID.sFeedBack=0;
    __GmWPID.sPreError=0;
    __GmWPID.sPreDerror=0;
    __GmWPID.fKp=__KP;
    __GmWPID.fKi=__KI;
    __GmWPID.fKd=__KD;
    __GmWPID.iPreU=0;
}
                Core function 2: __SPIDContr
/****************************************************************
** Function name:__SPIDContr
** Descriptions:Straight line PID control
** input parameters:None
** output parameters:None
** Returned value:None
****************************************************************/
void__SPIDContr(void)
```

```
    {
        float  error,d_error,dd_error;
        static uint8   K_I=1;
        error=__GmSPID.sRef-__GmSPID.sFeedBack; // Deviation calculation
        d_error=error-__GmSPID.sPreError;
        dd_error=d_error-__GmSPID.sPreDerror;
        if(error>Deadband)
          error-=Deadband;
        else if(error<-Deadband)
          error+=Deadband;
        else
          error=0;
        if((error>error_IMAX)||(error<-error_IMAX))
          K_I=0;
        else
          K_I=1;
        __GmSPID.sPreError=error; //Deviation storage
        __GmSPID.sPreDerror=d_error;
        __GmSPID.iPreU+=(int16)(__GmSPID.fKp*d_error+K_I*__GmSPID.
fKi*error+__GmSPID.fKd*dd_error);
    }
```

                  Core function 3: __WPIDContr

```
    /****************************************************************
    ** Function name:__WPIDContr
    ** Descriptions:Turning PID control
    ** input parameters:None
    ** output parameters:None
    ** Returned value:None
    ****************************************************************/
    void__WPIDContr(void)
    {
        float  error,d_error,dd_error;
        static uint8   K_I=1;
        error=__GmWPID.sRef+0.5*GsTpusle_T-__GmWPID.sFeedBack;
// Deviation calculation
        d_error=error-__GmWPID.sPreError;
        dd_error=d_error-__GmWPID.sPreDerror;
        if(error>Deadband)
          error-=Deadband;
        else if(error<-Deadband)
          error+=Deadband;
        else
          error=0;
        if((error>error_IMAX)||(error<-error_IMAX))
          K_I=0;
        else
```

```
            K_I=1;
            __GmWPID.sPreError=error;//Deviation storage
            __GmWPID.sPreDerror=d_error;
            __GmWPID.iPreU+=(int16)(__GmWPID.fKp*d_error+K_I*__GmWPID.
fKi*error+__GmWPID.fKd*dd_error);
        }
```

                        Core function 4: __PIDContr

```
/**************************************************************
** Function name:__PIDContr
** Descriptions:PID control,through pulses control motor
** input parameters:None
** output parameters:None
** Returned value:None
**************************************************************/
void __PIDContr(void)
{
    __SPIDContr();
    __WPIDContr();
    __GmLeft.sSpeed=__GmSPID.iPreU-__GmWPID.iPreU;
    if(__GmLeft.sSpeed>=0){
     __GmLeft.cDir=__MOTORGOAHEAD;
    if(__GmLeft.sSpeed>=U_MAX)    // Maximum overflow prevention
        __GmLeft.sSpeed=U_MAX;
    if(__GmLeft.sSpeed<=U_MIN)      // Minimum overflow prevention
        __GmLeft.sSpeed=U_MIN;
    }
    else{
      __GmLeft.cDir=__MOTORGOBACK;
      __GmLeft.sSpeed*=-1;
    if(__GmLeft.sSpeed>=U_MAX)
        __GmLeft.sSpeed=U_MAX;
    if(__GmLeft.sSpeed<=U_MIN)
        __GmLeft.sSpeed=U_MIN;
    }

    __GmRight.sSpeed=__GmSPID.iPreU+__GmWPID.iPreU;
    if(__GmRight.sSpeed>=0){
     __GmRight.cDir=__MOTORGOAHEAD;
    if(__GmRight.sSpeed>=U_MAX)
        __GmRight.sSpeed=U_MAX;
    if(__GmRight.sSpeed<=U_MIN)
        __GmRight.sSpeed=U_MIN;
    }
    else{
      __GmRight.cDir=__MOTORGOBACK;
      __GmRight.sSpeed*=-1;
```

```
    if(__GmRight.sSpeed>=U_MAX)
        __GmRight.sSpeed=U_MAX;
    if(__GmRight.sSpeed<=U_MIN)
        __GmRight.sSpeed=U_MIN;
    }
    __rightMotorContr();
    __leftMotorContr();
}
```

Flow chart. These two values, __GmWPID.sRef and __GmSPID.sRef, are modified to control Micromouse to move forward or turn at a steady speed. The flow chart is shown as Fig. 2–1–5.

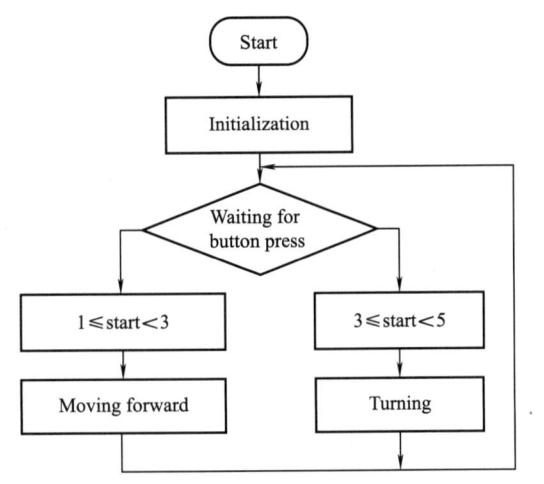

Fig. 2–1–5　Micromouse PID control flow chart

SysTick_Handler

```
void SysTick_Handler(void)
{
    __Encoder();  //Collection of motor pulse value and speed value
    __PIDContr(); //Using PID algorithm to control motor speed
}
```
main.c
```
/*********************************************************
** Function name:main
** Descriptions:main function
** input parameters:None
** output parameters:None
** Returned value:None
*********************************************************/
main (void)
```

```
{
    uint8 start=0;
    SystemInit();
    RCC_Init();
    JTAG_Set(1);
    MouseInit();
    PIDInit();
    ZLG7289Init();
    delay(100000);
    delay(100000);
    GPIO_Config1();
    USART1_Config1();
    NVIC_Config1();
    floodwall();
    GPIO_SetBits(GPIOB,GPIO_Pin_12);
    while (1) {
            if (startCheck()==true)
            {
                start++;
            }
            if((start<3)&&GucGoHead)
            {
__GmSPID.sRef=100;
__GmWPID.sRef=0;
}
if((start>=3)&&GucGoHead)
            {
__GmSPID.sRef=100;
__GmWPID.sRef=100;
}
        }
    }
```

# Task 2　Accurate Control of Micromouse Turning

In addition to going straight, Micromouse will encounter a lot of turns in the maze. Whether the turn is accurate or not has a great impact on the follow-up operation of Micromouse.

There are three common turning modes of Micromouse, as shown in Fig. 2–1–6.

The first and second methods are easy to control, the whole process are clear, and the stability is high, but the turning time is a little longer, which are

usually used by stepping motors; the third method is faster, but it is more difficult to control, which is usually adopted by DC motors.

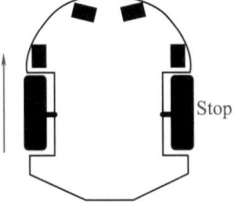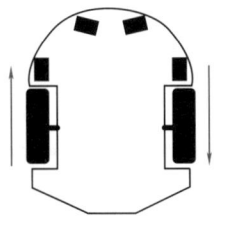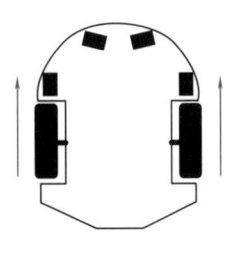

Fig. 2-1-6   Three turning modes

## Experiment   Optimal turning

When Micromouse turns in a maze, it needs to consider not only the turning angle but also the turning time. Turning too early or too late will affect the subsequent operation of Micromouse.

Micromouse enters the turning process at $t_0$, $t_0$ to $t_1$ is the pulse operation time before turning, $t_1$ to $t_2$ is the turning time, turning ends at $t_2$, and the pulse operation time after turning is $t_2$ to $t_3$, as shown in Fig. 2-1-7.

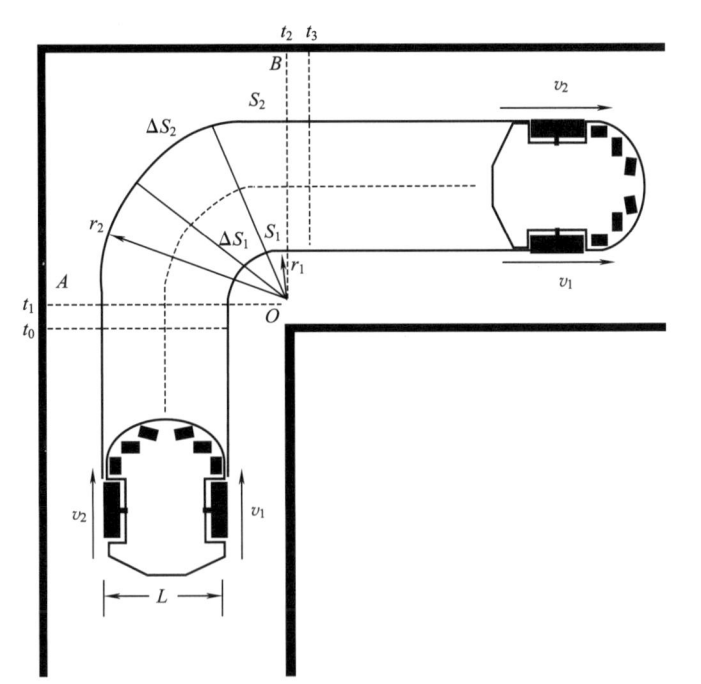

Fig. 2-1-7   The whole process of turning

Video

Experiment:
optimal
turning

## 1. Pulse-control before/after turning

If the pulse is too large before turning, it will be close to the outside wall after turning. If the pulse is too small, it will be close to the inner wall after turning, as shown in Fig. 2-1-8.

The pulses size after turning will affect the next turn.

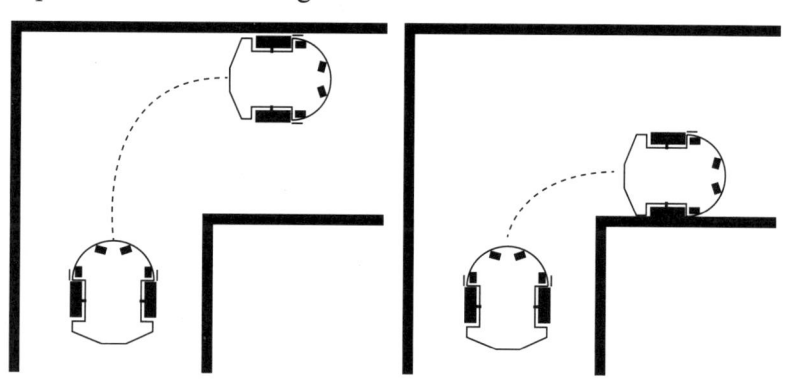

Fig. 2-1-8    Pulse-control before turning

Core function 1: mazeSearch

```
/*******************************************************************
** Function name:mazeSearch
** Descriptions:Search program
** input parameters:None
** output parameters:None
** Returned value:None
*******************************************************************/
staticint16   GuiTpusle_LR_l=0;
static int16  GuiTpusle_LR_r=0;
static uint8  sjjj;
void mazeSearch(void)
{
  :
  :
if (cL) {        /*Whether allowing detection on the left*/
            if  ((__GucDistance[__LEFT]&0x01)==0)
              {
/*If there is a wall on the left, jump out of the program*/
              __GmRight.uiPulse=__GmRight.uiPulseCtr+17800-
GuiTpusle_LR-GuiTpusle_LR_l;
              __GmLeft.uiPulse=__GmLeft.uiPulseCtr+17800-
GuiTpusle_LR-GuiTpusle_LR_l;
                          /* The pulses before left turning */
  :
  :
```

```
    if (cR) {           /*Whether allowing detection on the right*/
                if ((__GucDistance[__RIGHT]&0x01)==0){
    /*If there is a wall on the right, jump out of the program*/
                    __GmRight.uiPulse=__GmRight.uiPulseCtr+19800-
GuiTpusle_LR-GuiTpusle_LR_r;
                    __GmLeft.uiPulse=__GmLeft.uiPulseCtr+19800-
GuiTpusle_LR-GuiTpusle_LR_r;
    /*The pulses before right turning*/
        :
        :
        :
    }
    Core function 2: mouseTurnright(A part of the program)

/*************************************************************
** Function name:mouseTurnright
** Descriptions:Turn right
** input parameters:None
** output parameters:None
** Returned value:None
*************************************************************/
    void mouseTurnright(void)
    {
        :
        :
        __GmLeft.uiPulse=12000- piancha_r*2886.67;
        __GmLeft.uiPulseCtr=0;
        __GmRight.uiPulse =12000- piancha_r*2886.67;
//The pulses after turn right(the pulses after turn left is similar)
        __GmRight.uiPulseCtr=0;
        :
        :
        :
    }
```

## 2. Turning speed control

When Micromouse turns, the faster the straight speed is, the easier it is to slip; the faster the rotation speed, the smaller the turning radius. So many experiments are needed to find out the most perfect parameters.

Straight speed SPID=(V1+V2)/2.

Rotation speed WPID=(V1−V2)/2.

```
    Core function 3: mouseTurnright (The whole program)

/*************************************************************
** Function name:mouseTurnright
** Descriptions:Turn right
** input parameters:None
** output parameters:None
```

```
** Returned value:None
***************************************************************/
static uint8 after_turnright;
static uint8 aaaaa;
static uint8 lwt1;
float bc_r;
static int sssss=0;
void mouseTurnright(void)
{
static int ir=0;
float leftd=left_distance;
  piancha_r=pianchar(leftdis,leftd);
    GW=0;
    time=0;
    __GucMouseState=__TURNRIGHT;
    __GmRight.cState=__MOTORRUN;
    __GmLeft.cState=__MOTORRUN;
    GucMouseDir=(GucMouseDir+2)%8;      /*Direction mark*/
    __GmSPID.sRef=145;                  //Straight speed
    __GmWPID.sRef=-86;                  //Rotation speed
     while(1)
     {
         if(GW>150000)
         {
           break;
         }
     }
    ir++;
    __GmWPID.sRef=0;
    __GucMouseState=__GOAHEAD;
    GuiSpeedCtr=3;
    __GmLeft.uiPulse=12000-piancha_r*2886.67;
    __GmLeft.uiPulseCtr=0;
    __GmRight.uiPulse=12000-piancha_r*2886.67;
    __GmRight.uiPulseCtr=0;

    while ((__GmRight.uiPulseCtr+200)<=__GmRight.uiPulse);
    while ((__GmLeft.uiPulseCtr+200)<=__GmLeft.uiPulse);
    __GucMouseState=__TURNRIGHT;
    GuiSpeedCtr=__SPEEDUP;
     __GmRight.cState=__MOTORSTOP;
     __GmLeft.cState=__MOTORSTOP;
     __GmRight.sSpeed=0;
     __rightMotorContr();
     __GmLeft.sSpeed=0;
     __leftMotorContr();
     __GmRight.uiPulseCtr=0;
```

```
__GmLeft.uiPulseCtr=0;
mouseStop ( ) ;
while (1) ;
}
```

Flow chart. This experiment takes straight movement and right turn as an example to verify the whole turning process of Micromouse. The flow chart is shown as Fig. 2–1–9.

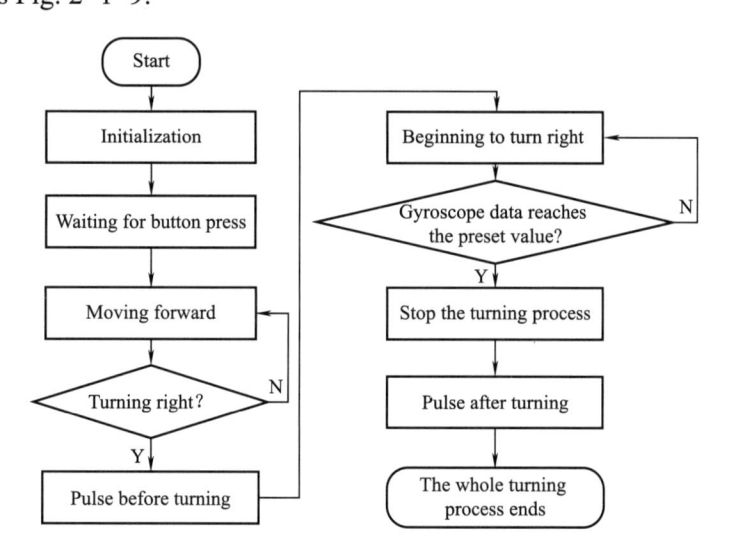

Fig. 2–1–9　Verification of Micromouse turning process flow chart

Main program:

```
                    main.c
/*****************************************************************
** Function name:main
** Descriptions:Main fanction
** input parameters:None
** output parameters:None
** Returned value:None
*****************************************************************/
main (void)
{
    uint8 n=0;   /*The number of coordinates with multiple branches*/
    uint8 ucRoadStat=0;
        /*Counting the number of directions can move forward*/
    uint8 ucTemp=0; /*Used for coordinate conversion in START state*/
    uint8 start=0;
    uint8 start_maxspeed=0;
    uint8 start_led=0;
```

```
SystemInit();
RCC_Init(),
JTAG_Set(1);
MouseInit();
PIDInit();
ZLG7289Init();
delay(100000);
//__ir_Get();
delay(100000);
GPIO_Config1();
USART1_Config1();
NVIC_Config1();
floodwall();
GPIO_SetBits(GPIOB,GPIO_Pin_12);
while (1) {
    switch (GucMouseTask) {   /*Processing of state machion*/
        case WAIT:
            sensorDebug();
            delay(10000);
          if (startCheck()==true)
            {
                start++;
            }
            if(start=1&&GucGoHead)
            {
                start=0;
              GucMouseTask=START;
              delay(1000000);
            }
                break;
        case START:
            mazeSearch();
            while (1)
            {
                if (startCheck()==true)
                {
                    break;
                }
            }
            break;
        default:
            break;
        }
    }
}
```

## Reflection and Summary

(1) What are the effects of P, I, D on the system of closed-loop control?

(2) How to control the turning angle of Micromouse?

(3) In order to realize the precise control of motor speed and turning angle, we introduce the PID control concept and the use of gyroscope. PID closed-loop control ensures the stability of speed and the gyroscope detects the deviation angle of Micromouse in real time, so as to achieve high-precision control of speed and posture.

<br>

## Project 2

## Intelligent Control Algorithm and Technology

### Learning objectives

(1) Understanding the principle of straight movement.

(2) Mastering the common way of recording the distance of Micromouse.

(3) Mastering the coordinate acquisition and suction control method of Micromouse.

Micromouse corrects the vehicle posture according to the sensor detection results, and judges the current coordinates according to the distance and turning direction. The application of suction cup enhances the friction of Micromouse.

## Task 1   Intelligent Obstacle Avoidance of Micromouse

The running process of Micromouse in a maze can be divided into straight movement and turning. Accurate infrared detection and turning parameters determine whether Micromouse can successfully find the destination through the maze. We have learned about infrared detection and accurate control of turning angle in the project 3 of chapter 1 and the project 1 of chapter 2 respectively. Next, we use the knowledge of these two aspects to realize the intelligent obstacle avoidance of Mircromouse.

### Experiment 1   Straight movement under correction

In the motor drive task, we have learned that there are no two motors with the same speed. In order to make Micromouse walk along the maze center line, we add encoders and infrared sensors. Encoders can record the motor speeds, but due to tire slipping and other reasons, infrared sensors are needed to detect

the distance between the two sides of the walls, judge whether Micromouse has offset, and correct for this situation.

```
                    Core function 1: SysTick_Handler
/*************************************************************
** Function name:SysTick_Handler
** Descriptions:Timing interrupt scan function
** input parameters:None
** output parameters:None
** Returned value:None
*************************************************************/
void SysTick_Handler(void)
{
  float leftbiao=left_distance;      // Calibration value
  float rightbiao=right_distance;
  float leftyu=left_yuzhi;
  float rightyu=right_yuzhi;
  static int8 n=0,m=0 ,k=0,l=0,a=0,b=0,c=0,w=0;
  static u16   t=0,s=0;
  static u32   encoder=0;
  uint16 Sp;
  if(zijiaozheng0==1)
  {
        if(zijiaozheng_flag)
        {
          sum_zijiaozheng+=Angle_TLY_Average;
          zijiaozheng_time++;
          if(zijiaozheng_time==3000){zijiaozheng=sum_
zijiaozheng/3000;zijiaozheng_flag=0;jiao=0;}
        }
  }
    __Encoder();
    TIM6_IRQHandler();
    Sp=__GmSPID.sFeedBack;
    encoder+=Sp;
    t++;
    s++;
    if(t==1000)
    {
        t=0;
        GPIO_SetBits(GPIOB, GPIO_Pin_12);
        if(s==2000)
        {
            s=0;
            GPIO_ResetBits(GPIOB, GPIO_Pin_12);
        }
```

```
                }
            switch (__GmRight.cState) {
        case__MOTORSTOP:
                __GmRight.uiPulse=0;
                __GmRight.uiPulseCtr=0;
                __GmLeft.uiPulse=0;
                __GmLeft.uiPulseCtr=0;
                break;
            case __WAITONESTEP:
                __GmRight.cState=__MOTORRUN;
                GsTpusle_T=dis(leftdis, rightdis, leftbiao, leftyu,
rightbiao, rightyu);
                break;
            case __MOTORRUN:                        /*Motor runs*/
              if (__GucMouseState==__GOAHEAD)
              /*Fine tune motor speed according to sensor states*/
                {    //Straight line correction of search and spurt
                    GsTpusle_T=disr(leftdis, rightdis, leftbiao,
leftyu, rightbiao, rightyu);
                    if(GuiSpeedCtr==__SPEEDUP)
                    {
                      k=(k+1)%5;//20
                      if(k==4)
                      __SpeedUp();
                    }
                    else if(GuiSpeedCtr==__SPEEDDOWN)
                    {
                        k=(k+1)%10;
                        if(k==5)
                        __SpeedDown();
                    }
                    else;
                }
                else
                {
                    GsTpusle_T=0;
                    voltageDetect();    //Reading gyro information
                }
                __PIDContr();
                break;
            case 4:
              GsTpusle_T=0;
              __PIDContr();
              break;

            default:
              break;
```

```
    }
}
```

Flow chart. In this experiment, sensors are used to detect the distance between the two sides of the walls to determine whether Micromouse has offset. The flow chart is shown as Fig. 2–2–1.

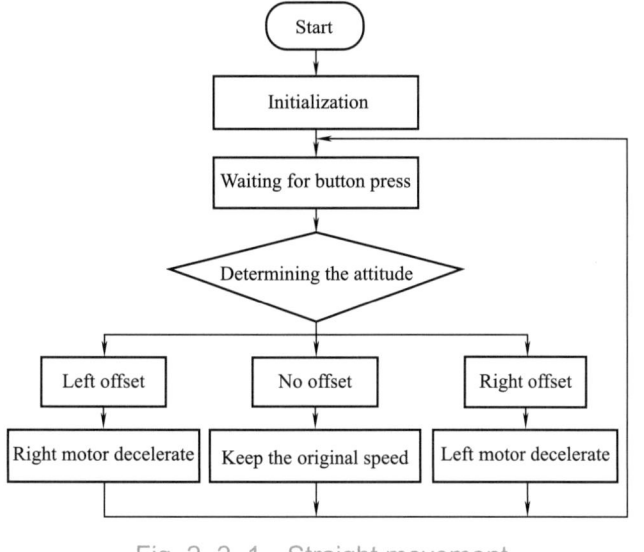

Fig. 2–2–1　Straight movement

Main program:

main.c

```
/**************************************************************
** Function name:main
** Descriptions:Main function
** input parameters:None
** output parameters:None
** Returned value:None
**************************************************************/
main (void)
{
    uint8 n=0;    /*The number of coordinates with multiple branches*/
    uint8 ucRoadStat=0;
        /*Counting the number of directions can move forward*/
    uint8 ucTemp=0;   /*Used for coordinate conversion in START state*/
    uint8 start=0;
    uint8 start_maxspeed=0;
    uint8 start_led=0;
    SystemInit();
```

```
RCC_Init();
JTAG_Set(1);
MouseInit();
PIDInit();
ZLG7289Init();
delay(100000);
delay(100000);
GPIO_Config1();
USART1_Config1();
NVIC_Config1();
floodwall();
GPIO_SetBits(GPIOB,GPIO_Pin_12);
while (1) {
    switch (GucMouseTask) {   /*Processing of state machine*/
        case WAIT:
            sensorDebug();
            delay(10000);

            if (startCheck()==true)
              {
                  start++;
               }
            if(start=1&&GucGoHead)
              {
                  start=0;
                  GucMouseTask=START;
                  delay(1000000);
              }
                break;
      case START:

          mazeSearch();

          while (1)
          {
              if (startCheck()==true)
              {
                  break;
              }
          }
          break;
      default:
          break;
      }
  }
}
```

## Experiment 2   Automatic turning

The premise of Micromouse turning automatically is that the infrared sensors detect the crossing. Combined with their own rules, choose the right turning direction.

```
              Core function 1: TIM6_IRQHandler
/***************************************************************
** Function name:TIM6_IRQHandler
** Descriptions:Timer function
** input parameters:None
** output parameters:None
** Returned value:None
***************************************************************/
void TIM6_IRQHandler(void)
{
    :
    :

            if(ucIRCheck[2]>GusDistance_L_Far)
            {
                __GucDistance[__LEFT]|=0x01;
            }
            else
            {
                __GucDistance[__LEFT]&=0xfe;
            }

            if(ucIRCheck[2]>GusDistance_L_Mid)
            {
                __GucDistance[__LEFT]|=0x02;
            }
            else
            {
                __GucDistance[__LEFT]&=0xfd;
            }

            if(ucIRCheck[2]>GusDistance_L_Near)
            {
                __GucDistance[__LEFT]|=0x04;
            }
            else
            {
                __GucDistance[__LEFT]&=0xfb;
            }
    :
    :
```

Core function 2: mazeSearch

```
/*****************************************************************
** Function name:mazeSearch
** Descriptions:Search function
** input parameters:None
** output parameters:None
** Returned value:None
*****************************************************************/
static int16   GuiTpusle_LR_l=0;
static int16   GuiTpusle_LR_r=0;
static uint8   sjjj;
void mazeSearch(void)
{
    int8 cL=0, cR=0, cCoor=1;
    if (__GmLeft.cState)
    {
        cCoor = 0;
    }
    if((__GucMouseState==__TURNRIGHT)||(__GucMouseState==__TURNLEFT))
    {
        __GmLeft.uiPulseCtr =40000;
        __GmRight.uiPulseCtr =40000;
        cL = 1;
        cR = 1;
        if(((__GucDistance[__FRONTR]!=0)&&(__GucDistance[__
FRONTL]!=0))||((__GucDistance[__LEFT]&0x01)==0)||((__
GucDistance[__RIGHT]&0x01) == 0))
            {
              if((__GucDistance[__FRONTR]!=0)&&(__GucDistance[__FRONTL]!=0))
              {
              GuiTpusle_LR =16000;
                if((__GucMouseState==__TURNRIGHT)&&((__
GucDistance[ __LEFT]&0x01) == 0)){GuiTpusle_LR_l=500;W_l=0;}
                  if((__GucMouseState==__TURNRIGHT)&&((__
GucDistance[__RIGHT]&0x01) == 0)){GuiTpusle_LR_r=0;W_r=-0;}
                    if((__GucMouseState==__TURNLEFT)&&((__
GucDistance[ __LEFT]&0x01) == 0)){GuiTpusle_LR_l=0;W_l=0;}
                    if((__GucMouseState==__TURNLEFT)&&((__
GucDistance[__RIGHT]&0x01) == 0)){GuiTpusle_LR_r=-4000;W_r=0;}
                }
                else
                  GuiTpusle_LR=12000;
            }
            else{
              GuiTpusle_LR=0;
            }
        }
```

```
        else{
          GuiTpusle_LR=0;
        }
        __GucMouseState=__GOAHEAD;
        __GiMaxSpeed=SEARCHSPEED;
        __GmRight.uiPulse=MAZETYPE * ONEBLOCK;
        __GmLeft.uiPulse=MAZETYPE * ONEBLOCK;
        __GmRight.cState=__MOTORRUN;
        __GmLeft.cState=__MOTORRUN;
         GuiSpeedCtr=__SPEEDUP;
        while (__GmLeft.cState !=__MOTORSTOP)
        {
            if (__GmLeft.uiPulseCtr>=ONEBLOCK)
            {              /*Judging whether complete one cell*/
               __GmLeft.uiPulse-=ONEBLOCK;
               __GmLeft.uiPulseCtr-=ONEBLOCK;
               if (cCoor)
               {
    if(((__GucDistance[__FRONTR]!=0)&&(__GucDistance[__FRONTL]!=0
))&&(ucIRCheck[2]>GusDistance_L_Far)&&(ucIRCheck[3]>GusDistance_R_
Far))//0x01
                 {
                   GucFrontNear=1;
                    goto End;
                 }
                  __mouseCoorUpdate();    /*Updating coordinate*/
               }
               else
               {
                   cCoor=1;
               }
            }
            if (__GmRight.uiPulseCtr>=ONEBLOCK) {
               __GmRight.uiPulse-=ONEBLOCK;
               __GmRight.uiPulseCtr-=ONEBLOCK;
            }
            if (cL) {    /*Whether allowing detection on the left*/

                if  ((__GucDistance[__LEFT]&0x01)==0)
                {
    /*If there is a branch on the left, jump out of the program*/
                   __GmRight.uiPulse=__GmRight.uiPulseCtr+17800-
GuiTpusle_LR-GuiTpusle_LR_l;
                   __GmLeft.uiPulse=__GmLeft.uiPulseCtr+17800-
GuiTpusle_LR-GuiTpusle_LR_l;
                   while ((__GucDistance[__LEFT]&0x01)==0)
                   {
```

```
                                     if ((__GmLeft.uiPulseCtr+400)>__GmLeft.uiPulse)
                                     {
                                         GucFrontNear=0;
                                         goto End;
                                     }
                                 }
                                 __GmRight.uiPulse=MAZETYPE*ONEBLOCK;
                                 __GmLeft.uiPulse=MAZETYPE*ONEBLOCK;
                                 GuiSpeedCtr=__SPEEDUP;
                             }
                     } else {
                         if (ucIRCheck[2]>GusDistance_L_Far) {
                             cL=1;
                         }
                     }
                     if (cR) {
                                 /*Whether allowing detection on the right*/
                         if ((__GucDistance[__RIGHT]&0x01)==0){//
mouseStop();while(1); /*If there is a branch on the right, jump
out of the program*/
                                 __GmRight.uiPulse=__GmRight.uiPulseCtr+19800-
GuiTpusle_LR-GuiTpusle_LR_r;
                                 __GmLeft.uiPulse=__GmLeft.uiPulseCtr+19800-
GuiTpusle_LR-GuiTpusle_LR_r;
                                 while ((__GucDistance[__RIGHT]&0x01)==0) {
                                     if ((__GmLeft.uiPulseCtr+400)>__GmLeft.uiPulse)
                                     {
                                         GucFrontNear=0;
                                         goto End;
                                     }
                                 }
                                 __GmRight.uiPulse=MAZETYPE*ONEBLOCK;
                                 __GmLeft.uiPulse=MAZETYPE*ONEBLOCK;
                                 GuiSpeedCtr=__SPEEDUP;
                             }
                     } else {
                         if (ucIRCheck[3]>GusDistance_R_Far)
                         {
                             cR=1;
                         }
                     }
                 }
        End:
                 __mouseCoorUpdate();   /*Updating coordinate*/
        }
```

Core function 3: crosswayChoice

```
/*************************************************************
** Functionname:crosswayChoice
** Descriptions:Selecting a branch as the forward direction
*************************************************************/
void crosswayChoice (void)
{
switch (SEARCHMETHOD) {
        //Judge the turning direction according to SEARCHMETHOD
        caseRIGHTMETHOD:
        mouseTurnright();        //Turning right
        break;
        caseLEFTMETHOD:
        mouseTurnleft();         //Turning left
        break;
        caseCENTRALMETHOD:
        centralMethod();
        break;
        caseFRONTRIGHTMETHOD:
        frontRightMethod();
        break;
        default:
        break;
    }
}
```

Flow chart. Micromouse can turn automatically according to the detection results of the sensor. Because there are many situations to consider when the Micromouse turning, this experiment takes only one intersection as an example to explain. The flow chart is shown as Fig. 2–2–2.

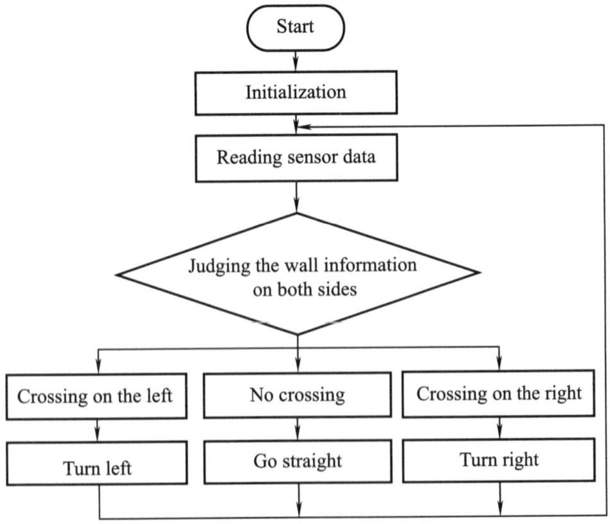

Fig. 2–2–2   Automatic turning of Micromouse

Main program:

```
                                main.c
/*****************************************************************
** Function name:main
** Descriptions:Automatic turning of Micromouse
** input parameters:None
** output parameters:None
** Returned value:None
*****************************************************************/
main (void)
{
    uint8 n=0;
      /*The number of coordinates with multiple branches*/
    uint8 ucRoadStat=0;
      /*Counting the number of directions can move forward*/
    uint8 ucTemp=0;
      /*Used for coordinate conversion in START state*/
    uint8 start=0;
    uint8 start_maxspeed=0;
    uint8 start_led=0;
    SystemInit();
    RCC_Init();
    JTAG_Set(1);
    MouseInit();
    PIDInit();
    ZLG7289Init();
    delay(100000);
    delay(100000);
    GPIO_Config1();
    USART1_Config1();
    NVIC_Config1();
    floodwall();
    GPIO_SetBits(GPIOB,GPIO_Pin_12);
    while (1) {
        switch (GucMouseTask) {  /*Processing of state machine*/
            case WAIT:
                sensorDebug();
                delay(10000);

                if (startCheck()==true)
                  {
                      start++;
                  }
                if(start=1&&GucGoHead)
                  {
                      start=0;
```

```
            GucMouseTask=START;
            delay(1000000);
        }
          break;
    case START:
        mazeSearch();
         while (1)
         {
             if (startCheck()==true)
             {
                 break;
             }
         }
         break;
    default:
         break;
    }
  }
}
```

# Task 2   Coordinate Recognition of Micromouse

Through the previous study, we already know how to control the movement of Micromouse. Micromouse needs to avoid obstacles and find the destination in a maze. So how does Micromouse determine its current position in a maze? There are 256 cells in the standard competition maze. How to accurately determine their own position, become a necessary condition for Micromouse to find the maze.

There are many ways for Micromouse to record their moving distance. The two commonly used methods are as follows:

1) Recording the number of column grooves

Each increase in the number of grooves indicates an increase in the number of walking cells. Because the column groove is only 3 mm, the change of sensor detection data is very little, so using this method requires very high sensor detection accuracy.

2) Record the distance of the motor

Micromouse records the number of pulses sent by the motor encoder, calculates the number of turns rotated, and then multiplies the wheel hub diameter

to get the number of cells that Micromouse walks. In comparison, this method is relatively simple and accurate. TQD-Micromouse-JM II uses this method to obtain the current position coordinates.

● Video

Experiment:
pulse control

## Experiment 1　Pulse control of Micromouse runs

### Core function 1: testEncoder

```
/***************************************************************
** Function name:testEncoder
** Descriptions:Encoder pulse count
** input parameters:None
** output parameters:None
** Returned value:None
***************************************************************/
void testEncoder(void)
{
    __GmLeft.uiPulseCtr=0;
    __GmRight.uiPulseCtr=0;
    __GucMouseState=__GOAHEAD;
    __GiMaxSpeed=SEARCHSPEED;
    __GmRight.uiPulse=5*50000;   //The pulses of the motor needs to run
    __GmLeft.uiPulse=5*50000;
    __GmRight.cState=__MOTORRUN;
    __GmLeft.cState=__MOTORRUN;
     GuiSpeedCtr=__SPEEDUP;
   while ((__GmRight.uiPulseCtr+200)<=__GmRight.uiPulse);
   while ((__GmLeft.uiPulseCtr+200)<=__GmLeft.uiPulse);
}
```

### Core function 2: mouseStop

```
/***************************************************************
** Function name:mouseStop
** Descriptions:Micromouse stop program
** input parameters:None
** output parameters:None
** Returned value:None
***************************************************************/
void mouseStop(void)
{
    __GmRight.cState=__MOTORSTOP;
    __GmLeft.cState=__MOTORSTOP;
    __GmRight.sSpeed=0;
    __rightMotorContr();
    __GmLeft.sSpeed=0;
    __leftMotorContr();
}
```

Flow chart. In this experiment, the number of cells are observed by modifying the pulse number of motor. As shown in Fig. 2–2–3.

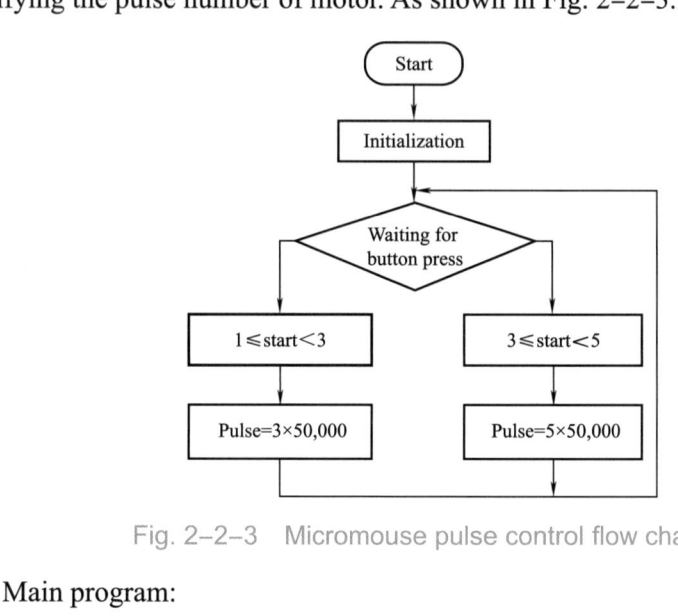

Fig. 2–2–3 Micromouse pulse control flow chart

Main program:

main.c

```
/************************************************************
** Function name:main
** Descriptions:Pulse control program
** input parameters:None
** output parameters:None
** Returned value:None
************************************************************/
main (void)
{
    uint8 start=0;
    SystemInit();
    RCC_Init();
    JTAG_Set(1);
    MouseInit();
    PIDInit();
    delay(100000);
    GPIO_Config1();
    while(1) {
        switch (GucMouseTask) {   /*Processing of state machine*/
            case WAIT:
                sensorDebug();
                delay(10000);
                if (startCheck()==true)
                {
```

```
                                         start++;
                                         delay(100);
                                       }
                                    if((start<3)&&(start>=1)&&GucGoHead)
                                    {
                                       start=0;
                                       GucMouseTask=START1;
                                       delay(1000000);
                                    }
                                 if((start<5)&&(start>=3)&&GucGoHead)
                                    {
                                       start=0;
                                       GucMouseTask=START2;
                                       delay(1000000);
                                    }
                                    break;
                          case START1:
                              testEncoder1();
                              mouseStop();
                              while (1)
                            {
                                 if (startCheck()==true)
                                 {
                                      break;
                                 }
                            }
                              break;
                          case START2:
                              testEncoder2();
                              mouseStop();
                               while(1)
                            {
                                 if (startCheck()==true)
                                 {
                                      break;
                                 }
                            }
                              break;
                          default:
                              break;
                      }
                  }
            }
```

We tried to modify __ GmRight.uiPulse and __ GmLeft.uiPulse many times.

It can be concluded that the number of pulses needed to run a single cell is about 51,600.

## Experiment 2    Coordinate acquisition of Micromouse

When Micromouse knows the number of cells it runs, combine with the number of turns and direction, it can get the current coordinates.

```
                    Core function: __mouseCoorUpdate

/*****************************************************************
** Functionname:__mouseCoorUpdate
** Descriptions:Updating the coordinate values according to
the current direction
** inputparameters:None
** outputparameters:None
** Returnedvalue:None
*****************************************************************/
void__mouseCoorUpdate (void)
{

    switch (GucMouseDir) {
    case 0:
        GmcMouse.cY++;
        break;
    case 2:
        GmcMouse.cX++;
        break;
    case 4:
        GmcMouse.cY--;
        break;
    case 6:
        GmcMouse.cX--;
        break;
    default:
        break;
    }
    __wallCheck();
    __mazeInfDebug();
}
```

Flow chart. In this experiment, the coordinates of the current position is determined by recording the number of pulses of the motor. As shown in Fig. 2-2-4.

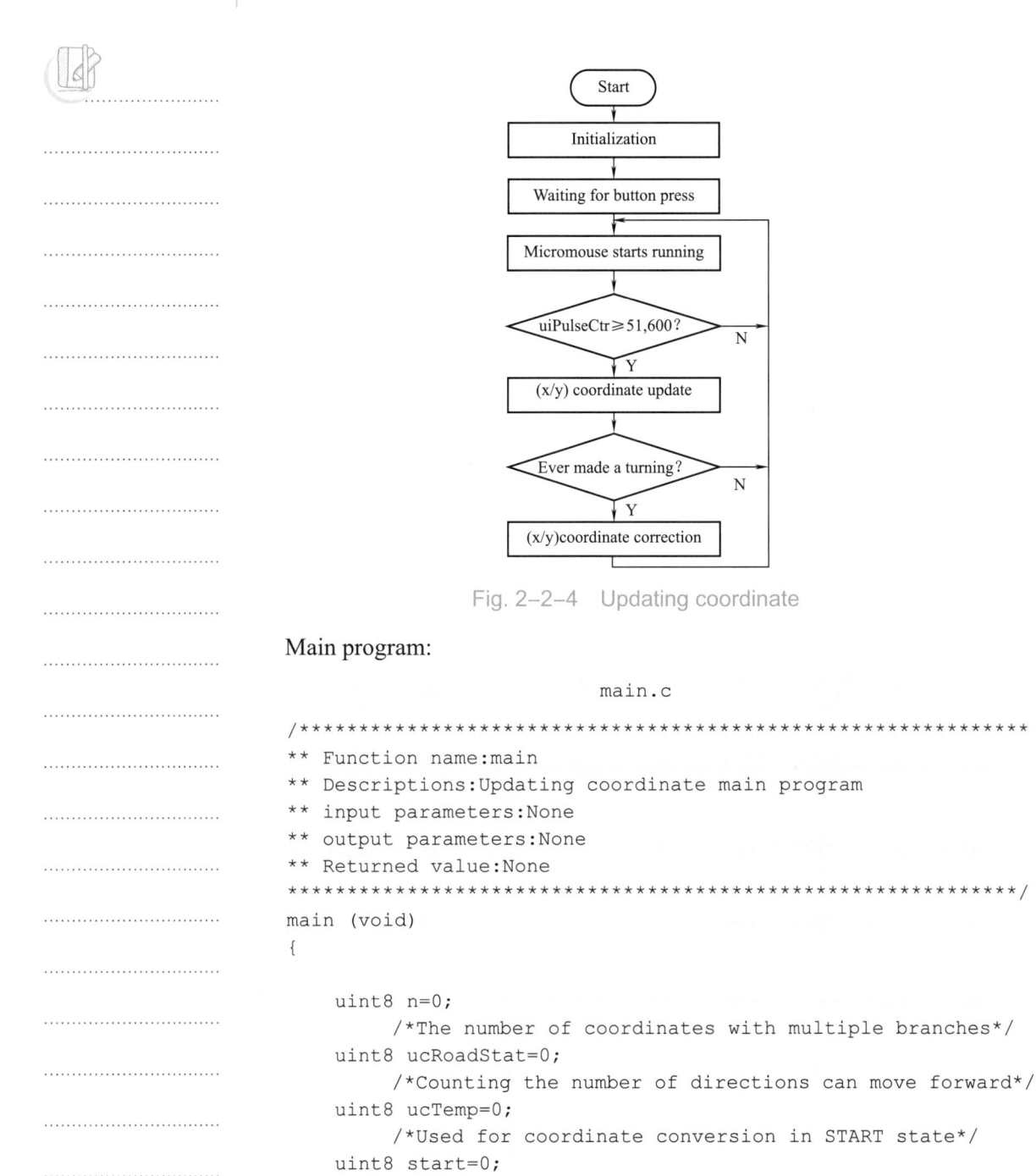

Fig. 2–2–4   Updating coordinate

## Main program:

```
                              main.c
/***************************************************************
** Function name:main
** Descriptions:Updating coordinate main program
** input parameters:None
** output parameters:None
** Returned value:None
***************************************************************/
main (void)
{

    uint8 n=0;
        /*The number of coordinates with multiple branches*/
    uint8 ucRoadStat=0;
        /*Counting the number of directions can move forward*/
    uint8 ucTemp=0;
        /*Used for coordinate conversion in START state*/
    uint8 start=0;
    uint8 start_maxspeed=0;
    uint8 start_led=0;
    SystemInit();
    RCC_Init();
    JTAG_Set(1);
```

```
    MouseInit();
    PIDInit();
    ZLG7289Init();
    delay(100000);
    delay(100000);
    GPIO_Config1();
    USART1_Config1();
    NVIC_Config1();
    floodwall();
    GPIO_SetBits(GPIOB,GPIO_Pin_12);
    while(1) {
        switch (GucMouseTask) {   /*Processing of state machine*/
            case WAIT:
                sensorDebug();
                delay(10000);

              if (startCheck()==true)
                {
                    start++;
                 }
                if(start=1&&GucGoHead)
                {
                    start=0;
                    GucMouseTask=START;
                    delay(1000000);
                }
                  break;
            case START:
            mazeSearch();
             while (1)
            {
                if (startCheck()==true)
                {
                    break;
                }
            }
            break;
            default:
            break;
        }
    }
}
```

# Task 3　Intelligent Suction Fan Technology

　　When Micromouse runs at high speed, it will slide laterally due to insufficient friction, so it needs to increase its friction on the maze floor. The suction fan is a good solution to this problem. It can extract the air between Micromouse and the maze floor, so that the internal air pressure is reduced, while the external air pressure is unchanged. Thus, Micromouse is pressed tightly on the maze floor under the external atmospheric pressure, making it more stable during acceleration and deceleration. The suction fan is shown in Fig. 2–2–5.

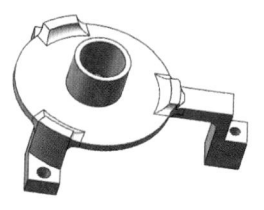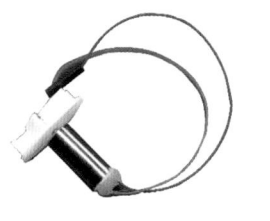

Fig. 2–2–5　The suction fan

## 1. Designing ideas and realizing methods of suction fan

　　TQD suction fan motor adopts special coreless DC motor with large torque and maximum speed of 50,000 r/min. The suction fan adopts MOSFET drive, PWM modulation, program modular design, easy debugging, and can configure the duty cycle according to the demand. According to the principle of aerodynamics, pump the bottom air out to form a relative vacuum environment, which can increase the friction between Micromouse and the floor and overcome the centrifugal force, as shown in Fig. 2–2–6. Under the action of the suction fan, reducing the center of gravity, turning at high speed without slowing down, more faster and stable. The sliding and shaking phenomena are effectively reduced.

● Video

Suction Fan
Technology

Fig. 2–2–6　Diagram of suction fan

The internal and external pressure is $P_I<P_O$ ($P_I$ is the internal pressure, $P_O$ is the external pressure), the internal and external pressure difference is $P=(P_O-P_I)=\dfrac{F}{S}$, and the friction force is $f=\mu N$. The unit of $P$ is Pa; the unit of $F$ is N; and the unit of $S$ is m$^2$. The schematic diagram of friction force and atmospheric pressure is shown in Fig. 2–2–7.

Fig. 2–2–7　Diagram of friction and atmospheric pressure

## 2. Suction fan motor drive and PWM speed regulation

The operation of Micromouse in the maze involves search, sprint, straight movement, and turning. Different operating states need different suction fan to increase friction. At the same time, in order to save battery power, it is necessary to use PWM technology to regulate motor speed. Suction fan power table is shown as Table 2–2–1.

Table 2–2–1　Suction fan power table

| PWM/% | Voltage/V | Current/mA | Power/W | Suction/g |
|---|---|---|---|---|
| 15 | 0.5 | 150 | 0.02 | 0 |
| 30 | 1.0 | 200 | 0.09 | 14.5 |
| 45 | 1.5 | 240 | 0.195 | 33.5 |
| 60 | 2.0 | 280 | 0.34 | 123.25 |
| 75 | 2.5 | 310 | 0.5 | 283.5 |
| 90 | 3.0 | 320 | 0.63 | 501.5 |

Core function: fan_init

```
/*************************************************************
** Function name:fan_init
** Descriptions:Suction fan initialization
** input parameters:Counter and divider
** output parameters:None
```

```
** Returned value:None
*****************************************************************/
void fan_init(u16 arr,u16 psc)
 {
    RCC->APB1ENR|=1<<2;       //TIM4 clock enable
    RCC->APB2ENR|=1<<3;       //PB clock enable
    RCC->APB2ENR|=1<<0;       //AFIO clock enable
    GPIOB->CRH&=0XFFFFFFF0;  //Reset PB8 state
    GPIOB->CRH|=0X0000000B;  //PB8 configuration
    AFIO->MAPR&=0XFFFFEFFF;  //Clear 12 bits of MAPR
    TIM4->CCR3=0;            //Initialize channel 3 duty cycle
    TIM4->ARR=arr;    //Set the auto reload value of the counter
    TIM4->PSC=psc;       //Prescaler does not divide frequency
    TIM4->CCMR2|=6<<4;       //CH3 PWM2 mode
    TIM4->CCMR2|=1<<3;       //CH3 load enable
    TIM4->CCER|=1<<8;        //CH3 output enable
    TIM4->CR1=0x0080;        //ARPE enable
    TIM4->CR1|=0x01;     //TIM4 enable
 }
```

When you need to call the suction fan, just add the code "TIM2 - > CCR3 = value;" and note that the value should not exceed 800, otherwise the circuit will be burned due to excessive current.

For example, to open the suction cup during a sprint:

```
case SPURT:
TIM2->CCR3=500;
mouseSpurt();
mouseTurnback();
objectGoTo(GucXStart,GucYStart);
```

## Reflection and Summary

(1) How to record the movement distance of Micromouse?

(2) How to control the strength of the suction fan?

(3) The suction fan technology is based on the fact that Micromouse is too fast, its weight is too light, and the friction is not enough when turning, which causes Micromouse to skid. Based on the principle of aerodynamics, the suction fan is used to pump the bottom air out, the pressure difference between the upper and lower is artificially created, thus enhancing the friction of Micromouse.

# Chapter **3**

# Advanced Skills and Competitions

The software and hardware of Micromouse and the basic programming and debugging methods have been introduced previously. This chapter mainly introduces the optimization algorithm according to the requirements of the IEEE International Standard Micromouse Competition. Mastering the specifications of competition enables the participants to complete the maze search and the best path selection at the fastest speed. Analyzing the key points of Micromouse competition cases, so as to prepare for participating in IEEE International Standard Micromouse competition.

TODO MICRO MOUSE I JQ J D J M

## ● ● ●   Project 1

### Acquisition and Storage of Maze Information

### Learning objectives

(1) Mastering the acquisition and storage methods of maze information of Micromouse.

(2) Mastering the path planning and decision algorithm of Micromouse.

(3) Mastering the principle of path planning of Micromouse.

How can Micromouse runs quickly in a maze? The main task of Micromouse is to complete the maze search and optimal path selection by IEEE International Standard rules. It is a competition that examines the ability of a system to detect, analyze, and make decisions about an unknown environment. Here is a brief introduction.

## Task 1   Conversion of Relative Direction to Absolute Direction

The maze is composed of $16 \times 16$ square cells with the size of 18 cm $\times$ 18 cm. In order to let Micromouse remember the information of each cell, we need to number the 256 cells. Obviously, it's very convenient to use coordinates.

We take the starting direction of Micromouse as a reference. At this time, the front is the positive direction of Y axis, the back is the negative direction of Y axis, the right side is the positive direction of X axis, the left side is the negatiue direction of X axis.

In order to convert the four direction parameters into symbols that can be recognized by the microcontroller, the upward direction is defined as 0, the right is 1, the bottom is 2 and the left is 3, as shown in Fig. 3–1–1.

With the coordinates and directions, Micromouse will know its position and direction at any time when walking in the maze. However, for Micromouse,

the directions of the infrared sensors are fixed, for the maze, the directions are changing with the direction of Micromouse. Because the reference is different. Therefore, this leads to two directions: relative direction and absolute direction.

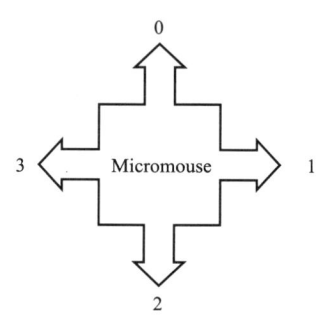

Fig. 3–1–1　Definition of direction value

Relative direction: the current walking direction of Micromouse is called the relative direction.

Absolute direction: the direction referenced to the coordinate plane of the maze is called the absolute direction.

How can the information stored by the sensors be processed more easily? Obviously, it's very convenient to store in an absolute direction. This involves the conversion of relative direction and absolute direction. In this chapter, the absolute direction value of Micromouse is recorded as Dir, and the relative direction of Micromouse is converted into absolute direction, as shown in Table 3–1–1.

Table 3–1–1　Relative direction converted to absolute direction

| Relative direction | Absolute direction |
| --- | --- |
| Front | Dir |
| Right | (Dir+1)%4 |
| Back | (Dir+2)%4 |
| Left | (Dir+3)%4 |

Sometimes the system also needs to find out the relative direction through the absolute direction. For example, to control Micromouse to turn to an absolute direction, it is necessary to calculate which relative direction the absolute direction is in, and Micromouse decides to turn by the relative direction.

First, according to the absolute direction of the target (Dir_dst) and the current absolute direction (Dir) to calculate the direction deviation (ΔDir), as shown in the following formula.

$$\Delta Dir = (Dir\_dst-Dir)\%4$$

At this time, the relative direction of Micromouse can be calculated by the direction deviation value, as shown in Table 3–1–2.

Table 3–1–2   Direction deviation converted to relative direction

| Direction deviation (ΔDir) | Relative direction |
| --- | --- |
| 0 | Front |
| 1 | Right |
| 2 | Back |
| 3 | Left |

Assuming that Micromouse has known the current coordinate($X$, $Y$), then the adjacent coordinate values in an absolute direction can be obtained, (The relative direction can be converted to absolute direction according to Table 3–1–1), as shown in the Table 3–1–3. The table is reversible, that is, the absolute direction can also be calculated according to the change of coordinate value.

Table 3–1–3   Coordinate conversion

| Absolute direction | Relative direction |
| --- | --- |
| Current position | ($X,Y$) |
| Up 0 | ($X,Y+1$) |
| Right 1 | ($X+1,Y$) |
| Down 2 | ($X,Y-1$) |
| Left 3 | ($X-1,Y$) |

# Task 2   The Information Storage Method of a Maze

For path planning, it is necessary to record the walls' information of all cells at first. Obviously, it is an effective method to establishing a two-dimensional array to define the coordinates of the entire maze. Each cell is defined as a coordinate, and the corresponding walls' information is stored in the established two-dimensional array.

When Micromouse reaches a cell, it should record the data of the wall according to the sensors detection results. In order to facilitate management and save storage space, the lower four bits of each byte variable are used to

store the wall data around a cell. The maze has $16 \times 16$ cells in total, so $16 \times 16$ two-dimensional array variables can be defined to save the whole maze wall's information, as shown in Fig. 3–1–2.

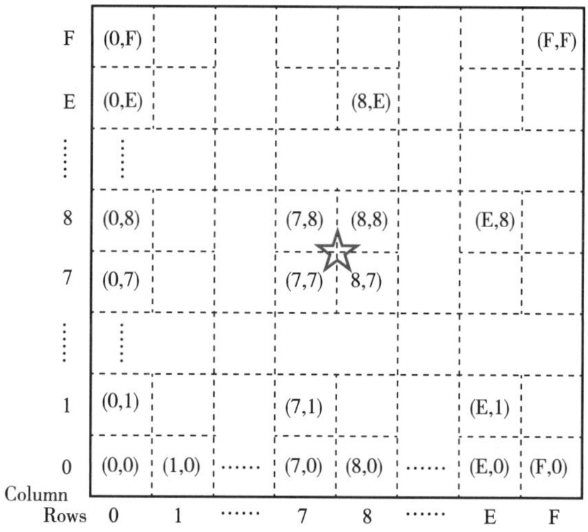

| | | |
|---|---|---|
| F | (0,F) | (F,F) |
| E | (0,E) | (8,E) |
| 8 | (0,8) | (7,8) (8,8) (E,8) |
| 7 | (0,7) | (7,7) 8,7 |
| 1 | (0,1) | (7,1) (E,1) |
| 0 | (0,0) (1,0) ⋯⋯ (7,0) (8,0) ⋯⋯ (E,0) (F,0) |
| Column Rows | 0 1 ⋯⋯ 7 8 ⋯⋯ E F |

Fig. 3–1–2   Maze coordinate definition

First, initialize the data of all the maze walls to 0. At least one side of a cell that Micromouse has walked through has no wall so that the data are not all 0; therefore, you can determine whether the cell has been searched by whether the data stored in the cell are 0. Wall information storage method is shown in Table 3–1–4.

Table 3–1–4   Wall information storage method

| Variable | Direction | One wall or not |
|---|---|---|
| bit0 | Up 0 | 1: No, 0: Yes |
| bit1 | Right 1 | 1: No, 0: Yes |
| bit2 | Down 2 | 1: No, 0: Yes |
| bit3 | Left 3 | 1: No, 0: Yes |
| bit7- bit4 | | Reserved |

## Reflection and Summary

(1) How does Micromouse record the walls' information in each cell?

(2) Micromouse needs to consider the current turning direction and absolute direction in operation.

# Project 2

## The Principle of Path Planning of Micromouse

### Learning objectives

(1) Understanding the common strategies of maze search.

(2) Understanding the principle of step map and turning weighted, and try to draw the optimal path on a competition maze.

## Task 1　Common Strategies of Maze Searching

Maze search method: without knowing the maze path, Micromouse must first explore all the cells in the maze until it reaches the destination. Micromouse going this procedure must know its position and posture at any time, and record whether there are walls around the cells. In order to save time during this process, we should try our best to avoid repeatedly searching the places.

Then how to explore the maze? Usually, there are two strategies: (1) Reaching the destination as soon as possible; (2) Searching the whole maze.

Both strategies have advantages and disadvantages. The first strategy can shorten the time needed to explore the maze, but we might not be able to obtain the data of the entire maze. If the path found is not the optimal path of the maze, which will affect the final sprint time of Micromouse. Using the second strategy can get the information of the entire maze so that we can find the optimal path. But, this strategy will take a long time to search.

## Task 2　The Basic Rules of Maze Searching

There are three common rules for searching: the right-hand rule, the left-

hand rule and the central rule, as shown in Fig. 3–2–1.

(a) The right-hand rule      (b) The left-hand rule      (c) The central rule

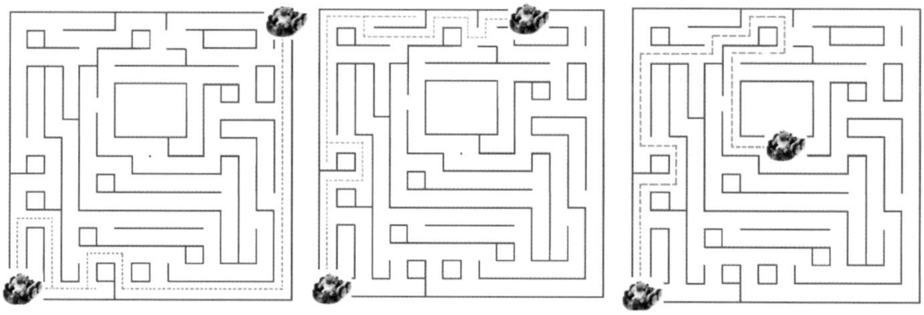

Fig. 3–2–1    The right-hand rule, the left-hand rule and the central rule

The right-hand rule: when there are multiple choices of heading, the order of priority is turning right, going straight, and turning left;

The left-hand rule: when there are multiple choices of heading, the order of priority is turning left, going straight, and turning right;

The central rule: when there are multiple choices of heading, the priority is turning towards the end.

Core function 1: rightMethod

```
/*************************************************************
** Function name:rightMethod
** Descriptions:Right-hand rule. The order of priority is turning
                right, going straight, and turning left;
** input parameters:None
** output parameters:None
** Returned value:None
*************************************************************/
void rightMethod (void)
{
    if ((GucMapBlock[GmcMouse.cX][GmcMouse.cY]&MOUSEWAY_R)&&
                            /*A path on the right*/
        (mazeBlockDataGet(MOUSERIGHT)==0x00)) {
                            /*Haven't crossed*/
        mouseTurnright();    /*Turning right*/
        return;
    }
    if ((GucMapBlock[GmcMouse.cX][GmcMouse.cY]&MOUSEWAY_F)&&
                            /*A path ahead*/
        (mazeBlockDataGet(MOUSEFRONT)==0x00)) {
                            /*Haven't crossed*/
        return;              /*Going straight*/
```

```
    }
    if ((GucMapBlock[GmcMouse.cX][GmcMouse.cY]&MOUSEWAY_L)&&
                            /*A path on the left*/
        (mazeBlockDataGet(MOUSELEFT )==0x00)) {
                            /*Haven't crossed*/
        mouseTurnleft();      /*Turning left*/
        return;
    }
}
```

### Core function 2: leftMethod

```
/*************************************************************
** Function name:leftMethod
** Descriptions:Left-hand rule. The order of priority is turning
left, going straight, and turning right;
** input parameters:None
** output parameters:None
** Returned value:None
*************************************************************/
void leftMethod (void)
{
  if ((GucMapBlock[GmcMouse.cX][GmcMouse.cY]&MOUSEWAY_L)&&
        /*There is a path on the left side of Micromouse*/
        (mazeBlockDataGet(MOUSELEFT)==0x00)) {
        /*The left side path of Micromouse didn't pass by*/
        mouseTurnleft();      /*Micromouse turns left*/
        return;
    }
    if ((GucMapBlock[GmcMouse.cX][GmcMouse.cY]&MOUSEWAY_F)&&
        /*There is a path on the front of Micromouse*/
        (mazeBlockDataGet(MOUSEFRONT)==0x00)) {
        /*The front of Micromouse didn't pass by*/
        return;             /*Micromouse doesn't need to turn*/
    }
    if ((GucMapBlock[GmcMouse.cX][GmcMouse.cY]&MOUSEWAY_R)&&
        /*There is a path on the right side of Micromouse*/
        (mazeBlockDataGet(MOUSERIGHT)==0x00)) {
        /*The right side path of Micromouse didn't pass by*/
        mouseTurnright(); /*Micromouse turns right*/
        return;
    }
}
```

### Core function 3: centralMethod

```
/*************************************************************
** Function name:centralMethod
** Descriptions:Central rule. The central rule determines which
search rule to use based on the current position of Micromouse in
```

```
the maze.
** input parameters:None
** output parameters:None
** Returned value:None
****************************************************************/
void centralMethod (void)
{
    if (GmcMouse.cX&0x08) {
        if (GmcMouse.cY&0x08) {
            /*
             * Micromouse is at the upper-right of the maze
             */
            switch (GucMouseDir) {
            case UP:    /*Currently, Micromouse is facing up*/
                leftMethod();          /*left-hand rule*/
                break;
            case RIGHT: /*Currently, Micromouse is facing right*/
                rightMethod();         /*right-hand rule*/
                break;
            case DOWN:  /*Currently, Micromouse is facing down*/
                frontRightMethod();  /*Center right rule*/
                break;
            case LEFT: /*Currently, Micromouse is facing left*/
                frontLeftMethod();   /*Center left rule*/
                break;
            default:
                break;              }
        } else {
            /*
             * Micromouse is at the lower-right of the maze
             */
            switch (GucMouseDir) {
            case UP:  /*Currently, Micromouse is facing up*/
                frontLeftMethod();   /*Center left rule*/
                break;
            case RIGHT: /*Currently, Micromouse is facing right*/
                leftMethod();          /*left-hand rule*/
                break;
            case DOWN: /*Currently, Micromouse is facing down*/
                rightMethod();         /*right-hand rule*/
                break;
            case LEFT: /*Currently, Micromouse is facing left*/
                frontRightMethod();  /*Center right rule*/
                break;
            default:
                break;            }
        }
```

```
    } else {
        if (GmcMouse.cY&0x08) {
            /*
             *  Micromouse is at the upper-left of the maze
             */
            switch (GucMouseDir) {
            case UP:     /*Currently, Micromouse is facing up*/
                rightMethod();        /*right-hand rule*/
                break;
            case RIGHT: /*Currently, Micromouse is facing right*/
                frontRightMethod();       /*Center right rule*/
                break;
            case DOWN: /*Currently, Micromouse is facing down*/
                frontLeftMethod();       /*Center left rule*/
                break;
            case LEFT: /*Currently, Micromouse is facing left*/
                leftMethod();            /*left-hand rule*/
                break;
            default:
                break;                 }
        } else {
            /*
             * Micromouse is at the lower-left of the maze
             */
            switch (GucMouseDir) {
            case UP:     /*Currently, Micromouse is facing up*/
                frontRightMethod();     /*Center right rule*/
                break;
            case RIGHT: /*Currently, Micromouse is facing right*/
                frontLeftMethod();      /*Center left rule*/
                break;
            case DOWN: /*Currently, Micromouse is facing down*/
                leftMethod();            /*left-hand rule*/
                break;
            case LEFT: /*Currently, Micromouse is facing left*/
                rightMethod();           /*right-hand rule*/
                break;
            default:
                break;                 }
        }
    }
}
```

# Task 3   The Step Map Making Method

Assuming that Micromouse has searched the whole maze or only part of the maze

including the start and the destination, and has recorded the walls data of each cell it has passed, then how can it find the optimal path from the start to the destination? The following introduces the concept and the method of step map making.

Step map is widely used in the field of geography and meteorology. It can mark the area range of the same altitude or the range and size of air pressure. Then the step map can be used on the maze map to calculate the distance between each maze cell and the destination, until get the distance between the start and the destination. By sorting the distance of each cell to the destination from large to small, we can find the optimal path in the maze.

Marking the starting as 1. According to the walls' information of each cell, the shortest steps to the start are marked on the cell so that the optimal path from any coordinate to the start is obtained, as shown in Fig. 3–2–2.

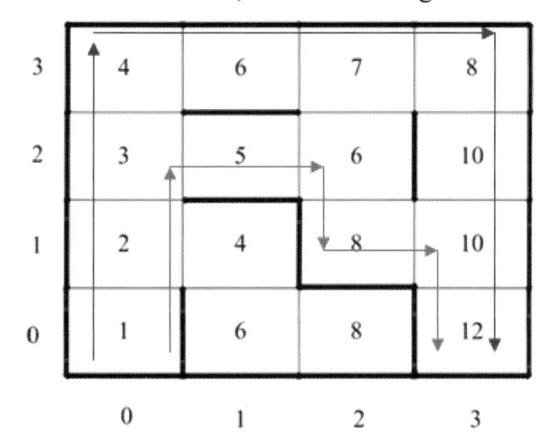

Fig. 3–2–2    The sketch map of step map

## Reflection and Summary

(1) What are the advantages and disadvantages of the left-hand rule, the right-hand rule and the central rule?

(2) The search rules of Micromouse are all composed of the left-hand rule and the righ-hand rule. According to certain rules, the maze is divided into several parts. Different search rules are selected when Micromouse has different positions and orientations.

(3) Micromouse needs to slow down and accelerate when it turns. So we need to weight the turn. Forward going have priority overturns, and the long straights have priority over short straights.

# Project 3

## Micromouse Program Design

### Learning objectives

Understanding the programming of Micromouse.

The flexibility and intelligence of TQD-Micromouse-JM II do not only depend on the structure and performance of hardware but also depend on the programming. The more intelligent is, the more complex the programming. In the Micromouse programming, the overall structure of the programming can be simply divided into two parts: bottom driver program and the top algorithm program.

The bottom driver program is mainly for the realization of some basic functions of Micromouse, such as controlling its going-forward $N$ cells in a straight line, measuring the distance it has moved forward, turning 90° to the right or left, preventing collision with the wall, and detecting the walls' information around the cell, etc.

The top algorithm program is mainly the intelligent algorithm of Micromouse, such as determining the action of Micromouse according to the maze information, remembering the map of the maze that it has walked, and finding the optimal path to the destination, etc.

## Task 1   Attitude Program Control

After the operation of TQD-Micromouse-JM II, it will process a lot of information and switch its states.

1) Waiting state

In this state, TQD-Micromouse-JM II stops at the start coordinate and waits for the start command. At the same time, the sensors' detection results and battery voltage are displayed in real time, which are convenient for debugging the

sensitivity of the sensors and replacing the battery. When the button that controls the start is pressed, TQD-Micromouse-JM II will enter the starting state.

2) Starting state

In this state, TQD-Micromouse-JM II will determine whether the start coordinate is (0,0) or (F,0) according to the direction of the first turning. Flow chart of the start coordinate judgment is shown as Fig. 3-3-1.

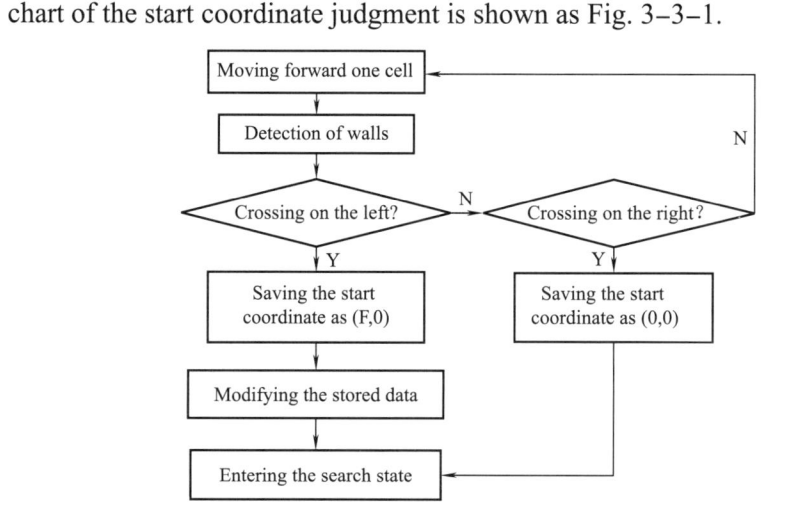

Fig. 3-3-1　Flow chart of the start coordinate judgment

3) Searching state

In this state, the task of TQD-Micromouse-JM II is to explore and memorize the maze map. Here we use the right-hand rule to search the whole maze. Flow chart of the maze searching is shown as Fig. 3-3-2.

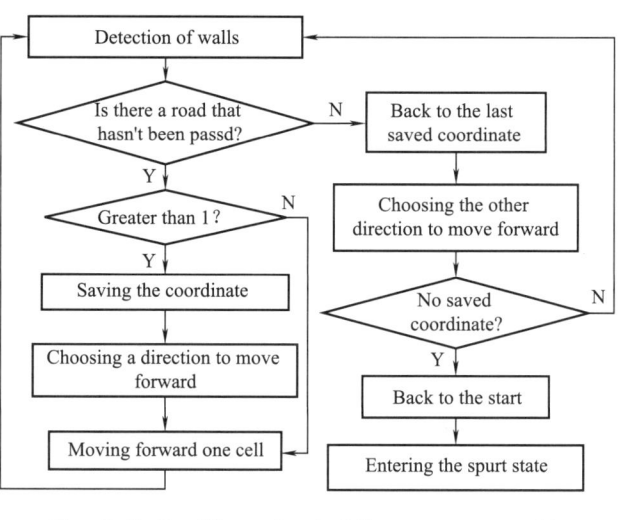

Fig. 3-3-2　Flow chart of the maze searching

4）Spurting State

After the maze search, Micromouse will find an optimal path to spurt to the destination according to the algorithm. It will return to the start after the spurt.

# Task 2   The Main Program Structures

Here are a few main functions that Micromouse calls when it runs.

1) Step map making program

Combining its own rules, Micromouse integrates the walls' information of all coordinates, and plans the optimal path.

```
                 Core function 1: mapStepEdit
/**************************************************************
** Function name:mapStepEdit
** Descriptions:Making step map with the start as the target point
** input parameters:uiX(The X-coordinate of the destination)
**                  uiY(The Y-coordinate of the destination)
** output parameters:GucMapStep[][](Step values of each coordinate)
** Returned value:None
**************************************************************/
void mapStepEdit (int8  cX, int8  cY)
{
    uint8 n=0;           /*Counting the number of crossings*/
    uint8 ucStep=1;      /*Step map value*/
    uint8 ucStat=0;
        /*Counting the number of directions can move forward*/
    uint8 i,j;

    GmcStack[n].cX=cX;   /*Saving start X-coordinate in stack*/
    GmcStack[n].cY=cY;   /*Saving start Y-coordinate*in stack*/
    n++;
    /*
     *  Step map value initialization
     */
    for (i=0;i<MAZETYPE;i++) {
        for (j=0; j<MAZETYPE;j++) {
            GucMapStep[i][j]=0xff;
        }
    }
  while (n) {
        GucMapStep[cX][cY]=ucStep++;  /*Filling in step value*/
/*Counting the direction that the current coordinate can move forward*/
        ucStat=0;
        if ((GucMapBlock[cX][cY]&0x01) &&  /*There is a way upper*/
```

```
                            (GucMapStep[cX][cY+1]>(ucStep))) {
      /*The upper step value is greater than the plan setting value*/
                    ucStat++;      /*Forward direction numbers plus 1*/
                }
                if ((GucMapBlock[cX][cY]&0x02) &&
      /*There is a way on the right*/
                    (GucMapStep[cX+1][cY]>(ucStep))) {
      /*The right step value is greater than the plan setting value*/
                    ucStat++;        /*Forward direction numbers plus1*/
                }
                if ((GucMapBlock[cX][cY]&0x04) &&
                    (GucMapStep[cX][cY-1]>(ucStep))) {
                    ucStat++;        /*Forward direction numbers plus1*/
                }
                if ((GucMapBlock[cX][cY]&0x08) &&
                    (GucMapStep[cX-1][cY]>(ucStep))) {
                    ucStat++;        /*Forward direction numbers plus1*/
                }
        /*If there is no direction to go forward, jumping to the nearest
saved branch point. Otherwise, selecting any one direction to go forward*/

                if (ucStat ==0) {
                    n--;
                    cX=GmcStack[n].cX;
                    cY=GmcStack[n].cY;
                    ucStep=GucMapStep[cX][cY];
                } else {
                    if (ucStat>1) {
      /*If there are multiple directions can move forward, save the coordinate*/
                        GmcStack[n].cX=cX; /*Saving X coordinate in stack*/
                        GmcStack[n].cY=cY; /*Saving Y coordinate in stack*/
                        n++;
                    }
                    /*
                     *  Choosing any one direction to move forward.
                     */

                    if ((GucMapBlock[cX][cY]&0x01) &&
                                            /*There is a way upper*/
                        (GucMapStep[cX][cY+1]>(ucStep))) {
      /*The upper step value is greater than the plan setting value*/
                        cY++;                    /*Modifying coordinate*/
                        continue;
                    }
                    if ((GucMapBlock[cX][cY]&0x02) &&
                                            /*There is a way on the right*/
```

```
                    (GucMapStep[cX+1][cY]>(ucStep))) {
/*The right step value is greater than the plan setting value*/
            cX++;                      /*Modifying coordinate*/
            continue;
        }
        if ((GucMapBlock[cX][cY]&0x04) &&
                                  /*There is a way lower*/
                    (GucMapStep[cX][cY-1]>(ucStep))) {
/*The lower step value is greater than the plan setting value*/
            cY--;                      /*Modifying coordinate*/
            continue;
        }
        if ((GucMapBlock[cX][cY]&0x08) &&
                                  /*There is a way on the left*/
                    (GucMapStep[cX-1][cY]>(ucStep))) {
/*The left step value is greater than the plan setting value*/
            cX--;              /*Modifying coordinate*/
            continue;
        }
    }
  }
}
```

2) Jumping to specified coordinate program

The purpose of this program block is making the Micromouse runs to the specified coordinate point by the shortest path. Of course, the premise is that Micromouse has searched this coordinate.

```
                Core function 2: objectGoTo
/***************************************************************
** Function name:objectGoTo
** Descriptions:Moving to the specified coordinate
** input parameters:cXdst(The X-coordinate of the destination)
**                  cYdst(The Y-coordinate of the destination)
** output parameters:None
** Returned value:None
***************************************************************/
void objectGoTo(int8  cXdst, int8  cYdst)
{
    uint8 ucStep=1;
    int8  cNBlock=0,cDirTemp;
    int8  cX,cY;
    GucCrossroad=0;
    cX=GmcMouse.cX;
    cY=GmcMouse.cY;
```

```
            mapStepEdit(cXdst,cYdst);                /*Making step map*/
        while ((cX !=cXdst) || (cY !=cYdst)) {
            ucStep=GucMapStep[cX][cY];
            /*
             *   Choosing one direction that the step map value is
smaller than the current one to move forward
             */
            if ((GucMapBlock[cX][cY]&0x01)&&   /*There is a way upper*/
                (GucMapStep[cX][cY+1]<ucStep)) {
                    /*The upper step value is a little smaller*/
                cDirTemp=UP;        /*Recording the direction*/
                if (cDirTemp==GucMouseDir) {
        /*Giving priority to the direction that do not need turning*/
                    cNBlock++;     /*Going ahead one cell*/
                    cY++;
                    if((GucMapBlock[cX][cY]&0x0f)==0x0f)
                      GucCrossroad++;
                    continue;        /*Skipping this loop*/
                }
            }
            if ((GucMapBlock[cX][cY]&0x02) &&
                                    /*There is a way on the right*/
                (GucMapStep[cX+1][cY]<ucStep)) {
                    /*The right step value is a little smaller*/
                cDirTemp=RIGHT;    /*Recording the direction*/
                if (cDirTemp==GucMouseDir) {
        /*Giving priority to the direction that do not need turning*/
                    cNBlock++;     /*Going ahead one cell*/
                    cX++;
                    if((GucMapBlock[cX][cY]&0x0f)==0x0f)
                      GucCrossroad++;
                    continue;        /*Skipping this loop*/
                }
            }
            if ((GucMapBlock[cX][cY]&0x04) &&
                                    /*There is a way lower*/
                (GucMapStep[cX][cY-1]<ucStep)) {
                    /*The lower step value is a little smaller*/
                cDirTemp=DOWN;          /*Recording the direction*/
                if (cDirTemp==GucMouseDir) {
        /*Giving priority to the direction that do not need turning*/
                    cNBlock++;          /*Going ahead one cell*/
                    cY--;
                    if((GucMapBlock[cX][cY]&0x0f)==0x0f)
                      GucCrossroad++;
                    continue;              /*Skipping this loop*/
                }
```

```
        }
        if ((GucMapBlock[cX][cY]&0x08) &&
                              /*There is a way on the left*/
            (GucMapStep[cX-1][cY]<ucStep)) {
                   /*The left step value is a little smaller*/
            cDirTemp=LEFT;        /*Recording the direction*/
            if (cDirTemp ==GucMouseDir) {
/*Giving priority to the direction that do not need turning*/
                cNBlock++;        /*Going ahead one cell*/
                cX--;
                if((GucMapBlock[cX][cY]&0x0f)==0x0f)
                  GucCrossroad++;
                continue;         /*Skipping this loop*/
            }
        }
        cDirTemp=(cDirTemp+8-GucMouseDir)%8;
                              /*Calculating direction offset*/
        GucDirTemp=cDirTemp;
        if (cNBlock) {
          if((GucCrossroad <=1)&&(cNBlock>1))
             mouseGoahead_L(cNBlock);
                                /*Going ahead cNBlock cells*/
          else{
            mouseGoahead_L(cNBlock);
            GucCrossroad=0;
          }
        }
        cNBlock=0;
/*Task reset. Controling Micromouse to turn.*
        switch (cDirTemp) {
        case 2:
            mouseTurnright_C();
            break;
        case 4:
            mouseTurnback();
            break;
        case 6:
            mouseTurnleft_C();
            break;
        default:
            break;
        }
    GmcMouse.cX=cX;
    GmcMouse.cY=cY;
    }
    /*
    *   Judging whether the task is completed.
```

```
                                */
                          if (cNBlock) {
                              if((GucCrossroad <=1)&&(cNBlock>1))
                                mouseGoahead_L(cNBlock);
                                                   /*Going ahead cNBlock cells*/
                              else{
                                mouseGoahead_L(cNBlock);
                                GucCrossroad=0;
                              }
                              GmcMouse.cX=cX;
                              GmcMouse.cY=cY;
                          }
                      }
```

3) Counting the number of branches not searched program

This program is used to count the total number of unsearched branches around the specified coordinate so that the system can use the searching strategy. The program is as below:

```
                       Core function 3: crosswayCheck
    /***************************************************************
    ** Function name:crosswayChe6ck
    ** Descriptions:Counting the number of directions that have not
    passed by.
    ** input parameters:ucX(The X-coordinate of the current point)
    **                   ucY(The Y-coordinate of the current point)
    ** output parameters:None
    ** Returned value:ucCt(The directions that have not passed by)
    ***************************************************************/
    uchar crosswayCheck(char  cX, char  cY)
    {
        uchar ucCt=0;
    if ((GucMapBlock[cX][cY]&0x01) &&
        /*Absolute direction, there's a way up here in the maze*/
        (GucMapBlock[cX][cY+1])==0x00)  /*Absolute direction,
    the upper path in the maze didn't pass by*/
            ucCt++;              /*Forward direction numbers plus 1*/
        }
        if ((GucMapBlock[cX][cY]&0x02)&&
     /*Absolute direction, there's a way on the right in the maze*/
            (GucMapBlock[cX+1][cY])==0x00) {
    /*Absolute direction, the right path in the maze didn't pass by*/
            ucCt++;              /*Forward direction numbers plus 1*/
        }
        if ((GucMapBlock[cX][cY]&0x04)&&
    /*Absolute direction, there's a way down here in the maze*/
```

```
        (GucMapBlock[cX][cY-1])==0x00) {
/*Absolute direction, the lower path in the maze didn't pass by*/
        ucCt++;              /*Forward direction numbers plus 1*/
    }
    if ((GucMapBlock[cX][cY]&0x08) &&
/*Absolute direction, there's a way on the left in the maze*/
        (GucMapBlock[cX-1][cY])==0x00) {
/*Absolute direction, the left path in the maze didn't pass by*/
        ucCt++;              /*Forward direction numbers plus 1*/
    }
    return ucCt;
}
```

## 4) TQD-Micromouse-JM II main Program

```
/********************************************************
** Function name:main
** Descriptions:Main function
** input parameters:None
** output parameters:None
** Returned value:None
********************************************************/
main (void)
{
    uint8 n=0;
        /*The number of coordinates with multiple branches*/
    uint8 ucRoadStat=0;
        /*Counting the number of directions can move forward*/
    uint8 ucTemp=0;
        /*Used for coordinate conversion in START state*/
    uint8 start=0;
    uint8 start_maxspeed=0;
    uint8 start_led=0;
    SystemInit();
    RCC_Init();
    JTAG_Set(1);
    MouseInit();
    PIDInit();
    ZLG7289Init();
    delay(100000);
    delay(100000);
    GPIO_Config1();
    USART1_Config1();
    NVIC_Config1();
    floodwall();
    GPIO_SetBits(GPIOB,GPIO_Pin_12);
    while (1) {
        switch (GucMouseTask) { /*Processing of state machine*/
```

```
                            case WAIT:
                              sensorDebug();
                              delay(10000);
                           if (startCheck()==true)
                              {
                                 start++;
                               }
                              if((start==1)&&GucGoHead)//&&GucGoHead
                              {
                                 start=0;
                                 zijiaozheng0=1;
                                  GPIO_ResetBits(GPIOB,GPIO_Pin_12);
                       delay(50000000);
                       GucMouseTask=START;
                                 delay(1000000);
                                 GPIO_SetBits(GPIOB,GPIO_Pin_12);
                                 zijiaozheng0=1;
                              }
                              if((start<5)&&(start>=3)&&GucGoHead)
                              {
                                GPIO_ResetBits(GPIOB,GPIO_Pin_12);
                                zijiaozheng0=1;
                                 start=0;
                            wallget();
                            delay(10000000);
                            StartGet();
                            delay(10000000);
                            delay(1000000);
                            delay(1000000);delay(1000000); delay(1000000);
                            GucMouseTask=SPURTL;
                                delay(1000000);
                              }
                                break;
                        case START:
             /*Judging the x coordinate of the start point of Micromouse*/
                     GPIO_ResetBits(GPIOB,GPIO_Pin_12);
                     mazeSearch();                      /*Searching forward*/
                          if (GucMapBlock[GmcMouse.cX][GmcMouse.cY]&0x08)
                          {
                              if (MAZETYPE==16)
                              {
                                  GucXGoal0=8;
                                  GucXGoal1=7;
                              }
                              GucXStart=MAZETYPE-1;
                              GmcMouse.cX=MAZETYPE-1;
                              ucTemp=GmcMouse.cY;
```

```
          do {
                  GucMapBlock[MAZETYPE-1][ucTemp]=Guc
                                   MapBlock[0][ucTemp];
                  GucMapBlock0[MAZETYPE-1][ucTemp]=Guc
                                   MapBlock0[0][ucTemp];
                  GucMapBlock0[0][ucTemp]=0;
                  GucMapBlock1[MAZETYPE-1][ucTemp]=Guc
                                   MapBlock1[0][ucTemp];
                  if(ucTemp>0)
                  {
                      GucMapBlock1[MAZETYPE-2][ucTemp-1]
                                              = 0x1d;
                  }
                  GucMapBlock1[0][ucTemp+1]=0x17;
                  GucMapBlock1[1][ucTemp]=0x1f;
                  GucMapBlock[0][ucTemp]=0x10;
                  GucMapBlock[1][ucTemp]=0x10;
              }while (ucTemp--);
          GucMapBlock1[0][0]=0x13;
          GucMapBlock1[1][0]=0x1b;
          GucMapBlock1[MAZETYPE-2][0]=0x19;
          /*Saving the start coordinates in OFFSHOOT[0]*/
          GmcCrossway[n].cX=MAZETYPE-1;
          GmcCrossway[n].cY=0;
          n++;
          GucMouseTask=MAZESEARCH;
      }
      if (GucMapBlock[GmcMouse.cX][GmcMouse.cY]&0x02)
      {
          GmcCrossway[n].cX=0;
          GmcCrossway[n].cY=0;
          n++;
          GucMouseTask=MAZESEARCH;
      }
    break;
case MAZESEARCH:
    centralMethodnew();
    goalwall();
    StartSave();
    wallsave();
    mouseTurnback();
    objectGoTo1(GucXStart,GucYStart);
    onestep3();
    mouseTurnbackqi();
    GucMouseTask=SPURTL;
    break;
case  SPURTL:
```

```
                                TIM4->CCR3=600;delay(1000000); delay(1000000);
delay(1000000);                             //Seting duty cycle and delay
                  StartGet();
                  wallget();
                  goalwall();
                  mouseSpurt_CC();
                  onestep();
                  mouseTurnback();
                  objectGoTo1(GucXStart,GucYStart);
                  onestep();
                  mouseTurnbackqidian();
                GucMouseTask=SPURT45;
                break;
              case SPURT45:
                  TIM4->CCR3=600;
                  mouseSpurt_45();
                  usepiancha=1;
                  onestep1();
                  mouseTurnback();
                  __GmSPID.sRef=145;
                  TIM4->CCR3=0;
                  objectGoTo1(GucXStart,GucYStart);
                  mouseTurnbackqi();
                  while (1)
                  {
                      if (startCheck()==true)
                      {
                          break;
                      }
                  }
                break;
              default:
                  break;
              }
          }
      }
```

## **Reflection and Summary**

(1) What are the relationships among the various stages of Micromouse programming?

(2) What are the main functions of Micromouse used for?

(3) The Micromouse program is executed by several main functions working together. Only when all the functions are debugged accurately, Micromouse can successfully reach the destination through the maze.

Appendix

MICROMOUSE JQ/JDJM

# Appendix A

## Micromouse Competition Going Popular in the World

Since 2019, it has been the most prosperous and fruitful period in the history of Micromouse competition.The International Micromouse Competition is held around the world, as shown in Fig.A–1.

In January, the International Micromouse Competition was held in Bombay, India.

In March, the APEC International Micromouse Competition was held in California, US.

In April, the International Micromouse Competition was held in Gondomar, Portugal.

In May, the IEEE Micromouse International Invitational Competition was held in Tianjin, China.

In June, the International Micromouse Competition was held in London, UK.

In August, the International Micromouse Competition was held in Chile.

In October, the International Micromouse Competition was held in Egypt.

In November, All Japan Micromouse International Competition was held in Tokyo, Japan.

Micromouse Competition Going Popular in the World

The International Micromouse Competition will become a booster of global higher education, vocational education, general education and technological innovation and integrated development of production and education. With the rapid development of artificial intelligence Micromouse competition, the education field timely introduced international well-known competitions to improve students' professional comprehensive ability, master the experience of practice and innovation, and help the integrated development of industry and

education, and cultivate more excellent seed talents for the industry, profession and enterprise.

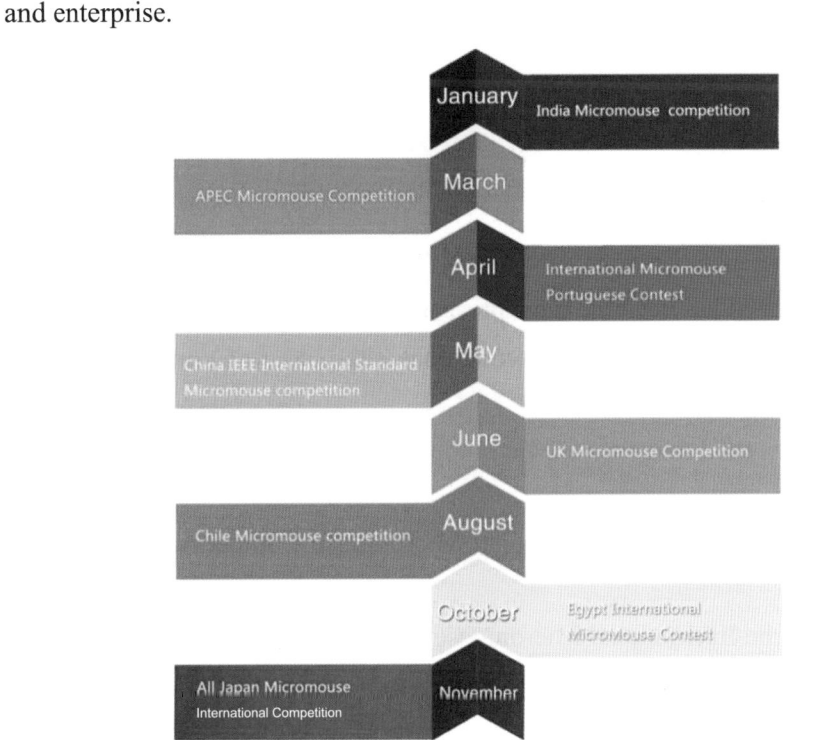

Fig. A–1　International Micromouse Competitions

1) China IEEE Micromouse International Invitational Competition

In 2009, Tianjin Qicheng Science and Technology Co., Ltd. introduced Micromouse competition into China, and carried out localized innovation and reform in the IEEE International Standard Micromouse Competition, which played a leading role in satisfying industrial optimization and upgrading, broadening international vision, gaining practice and innovation experience, and cultivating high-tech talents.

From 2016 to 2019, IEEE Micromouse International Invitational Competition has been successfully held for four times. The competition is hosted by Tianjin Municipal Education Commission and organized by Tianjin Qicheng Science and Technology Co. Ltd. and Tianjin Bohai Vocational Technical College, as shown in Fig. A–2.

Fig. A–2   The IEEE Micromouse International Invitational Competition

in China has been held since 2016

At present, IEEE Micromouse International Invitational Competition in China has set up "middle school, vocational college, bachelor's degree, master's degree and occupation" five competition group. It aims to improve the social participation and professional coverage of the competition. Micromouse has become an important carrier of systematic training and education. It fully embodies the combination of optical and electrical, software and hardware, control and machinery. While deducing the concept of "engineering" course, it extends and expands the concept of "innovative" course, which makes the content of students' learning and the teaching method of teachers have a new connotation, and truly focuses on the cultivation of comprehensive quality to create happy quality education.

IEEE Micromouse International Invitational Competition in China has the following characteristics.

(1) Participants: Facing not only college students, but also primary school, middle school and vocational workers, reflecting the characteristics of through training and lifelong education. It also includes international Micromouse professional players and previous international Micromouse competition winners.

(2) Maze site: There are 8×8 Micromouse maze sites for primary and secondary schools, and also have 16×16 full size classical Micromouse maze site for colleges and universities. Other than that, there is also a 25×32 half size Micromouse maze for elite players. Reflecting the extensibility of the competition, it takes the Micromouse competition as the core form, and students from different learning stages can participate in the competition.

(3) Competition events: There are not only Micromouse competition,

Video

China IEEE
Micromouse
International
Invitational
Competition

but also Robotracer race, which reflects the competition is both technical and engineering, and show the idea of engineering application oriented competition.

(4) Competition rules: A comparison of the similarities and differences in competition rules of general education, vocational education, higher education, vocational elite, as shown in Table A–1.

Table A–1　A comparison of similarities and differences in competition rules

| Entry category | General education | Vocational education | Higher education | Vocational elite |
|---|---|---|---|---|
| Competition form | (1) Online debugging of APP.<br>(2) Graphical programming.<br>(3) Application of IOT intelligent sensing technology.<br>(4) 8×8 maze race | (1) The oretical knowledge assessment.<br>(2) According to the referee the task programming and implement the corresponding function<br>(3) On-site technical defense.<br>(4) 16×16 classical maze racing | (1) DIY appearance and structure mechanical design.<br>(2) Hardware technology innovation.<br>(3) Program algorithm innovation.<br>(4) 16 × 16 classical maze racing | (1) DIY appearance and structure mechanical design.<br>(2) Hardware technology innovation.<br>(3) Program algorithm innovation.<br>(4) 25×32 half size maze racing |
| Competition content | (1) Assembly task 10%.<br>(2) Debugging task 40%.<br>(3) Racing task 50% | (1) Theoretical assessment 20%.<br>(2) Innovation 30%.<br>(3) Speed race 50% | (1) Innovation 20%.<br>(2) Speed race 80% | Speed race 100% |

International Micromouse experts on-site training guidance is shown in Fig. A–3.

Fig. A–3　International Micromouse experts on-site training guidance

2) US APEC International Micromouse Competition

In 1977, the first exciting Micromouse competition was held in New York, US. It was co-sponsored by IEEE and APEC. Thus, the most influential international American APEC world Micromouse Competition was born. Known as one of the world's three major Micromouse competitions, it has held 34 competitions until 2019.

The official website of APEC: http://www.apec-conf.org/.

American Micromouse enthusiasts website: http://Micromouse usa.com/, as shown in Fig.A–4.

 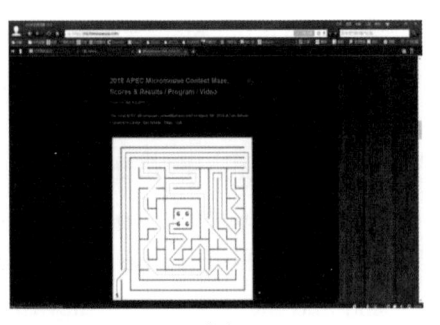

(a)                                                    (b)

Fig. A–4   Screenshots of official website of American APEC International Micromouse Competition

Competition time: Between February and April every year.

Competition venue: Varies annually (previous venues include North Carolina, Texas, Florida, California, etc.). Every year, different countries such as the United States, the United Kingdom, Japan, South Korea, Singapore, India, China have actively participated in the competition, as shown in Fig. A–5.

Fig. A–5   The 30th American APEC International Micromouse Competition

3) UK International Micromouse Competition

Since 1980, UK International Micromouse Competition has become one of the internationally well-known Micromouse competitions.

Competition time: June every year.

Competition venue: Birmingham City University.

The competition is sponsored by UK Micromouse and Robotics Society. The

characteristic of the Micromouse competition in UK is that it does not restrict anyone to participate,whether you are from middle school, university or social personnel. All the players are divided into different groups, and the difficulty of maze is adjusted appropriately. The competition is divided into line follower, wall follower, maze solver and other projects, attracting more than 50 teams from more than 10 countries of the world.

UK Micromouse Competition scoring rules: In the 16×16 maze, the participating Micromouse need to complete the search from the beginning to the end and the traversal of the whole maze, solve the best route and complete the sprint from the beginning to the end. Scoring time = search time (time used to find the end for the first time) / 30 + sprint time (high-speed sprint with the shortest path from the starting point to the end) + penalty time (3s/time for knocking into the wall).

The official website: https://ukmars.org/index.php/Main_Page, as shown in Fig. A–6.

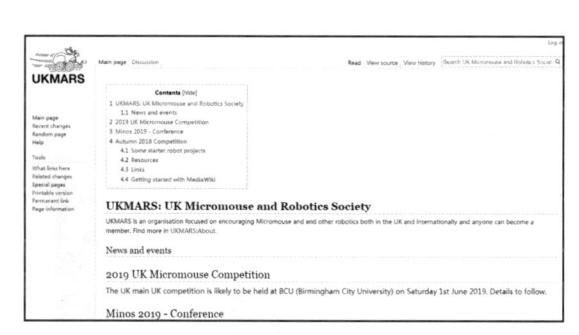

Fig. A–6    Screenshot of official website of UK Micromouse Competition

4) All Japan Micromouse International Competition

All Japan Micromouse International Competition has been held 40 times from 1980 to 2019.

Competition time: November or December every year.

Competition venue: Tokyo, Japan.

The official website: http://www.ntf.or.jp/mouse/Micromouse 2018/index. html, as shown in Fig. A–7.

Every year, Micromouse teams from more than 20 countries such as the United States, Britain, Japan, Singapore, China, Mongolia, Chile, Portugal participate in the competition, as shown in Fig. A–8.

● Video

All Japan
Micromouse
International
Competition

Fig. A–7    Screenshot of official website of All Japan Micromouse International Competition

Fig. A–8    Prize-giving of the 39th All Japan Micromouse International Competition

The competition consists of classic Micromouse event, half size Micromouse event and Robotrace event. The team consists of middle school students, college students and vocational elites. According to statistics, there are more than 300 teams. All Japan Micromouse International Competition can be said to represent today's international Micromouse technology field with the highest level and the strongest technology, so it has attracted much attention.

5) Chile Micromouse International Competition

The Ministry of Foreign Affairs of Chile hopes to promote the technological innovation of Chilean youth and international technological innovation, exchange and cooperation through Micromouse international competition, so as to promote the economic development of Chile. On December 3rd 2018, during the All Japan Micromouse International Competition, the Embassy of Chile in Japan hosted the "Chile International Micromouse Competition Seminar" and specially invited international experts (David Otten of the United States, Peter Harrison of the United Kingdom , Yukiko Nakagawa of Japan, Song Lihong of China, Benjamin of Chile,

etc.) jointly discussed the unified standards and specifications for the Micromouse maze international competition in Chile, as shown in Fig. A–9 and Fig. A–10.

Fig. A–9　Meeting of ministry of foreign affairs, Chile—discussion on the development of Micromouse

Fig. A–10　Chile Micromouse international competition seminar

● Video

Portuguese
Micromouse
International
Competition

6) Portuguese Micromouse International Competition

Portuguese Micromouse International Competition has helded in Gondomar on April 27, 2019, sponsored by University of Tras-os-Montes and Alto Douro, Portugal and the Technical Executive Committee.

Portuguese Micromouse International Competition, which started in 2011, aims to provide a complete technological learning environment through the cultivation of creativity and ability, and has been successfully held for 9 times.

Competition time: Every April/May.

Competition venue: Portugal.

The official website: https://www.micromouse.utad.pt/, as shown in Fig. A–11.

On April 27, 2019, at 18:00 local time, in Gondomar Coliseumm, Portual, teams from UK, China, Portugal, Spain, Brazil, Singapore and other countries, were competing a tense international Micromouse maze competition. With

the steady search and fast sprint of Chinese Micromouse, applause and cheers thundered in Gondomar Coliseumm...Micromouse of Qicheng achieved breakthrough results and won the second place in the world, as shown in Fig. A–12.

Fig. A–11　Screenshot of official website of Portuguese contest Micromouse International Competiton

Fig. A–12　Micromouse of Qicheng won the world second place in the Portugal competition

Antonio Valente, chairman of Micromouse Portuguese Contest Organizing Committee, said after the contest that in recent years, China's comprehensive national strength and technical strength have been increasing, especially in the field of education, more and more attention has been paid to technological innovation and engineering literacy. TQD-Micromouse participated in the Portugal international competition for the first time, and its excellent results are very gratifying, as shown in Fig. A–13.

Fig. A–13　On site technical exchange between Chinese and Portuguese Micromouse experts

7) India Interantional Micromouse Challenge

On January 4, 2020, the First International Micromouse Challenge of the 23rd Edition of Asia's Science and Technology Festival—Techfest 2020 was held in Mumbai, India. The delegations from India, China, Australia, Nepal, Sri Lanka, Bangladesh and other countries participated in the competition (see Fig. A–14). Micromouse team of Tianjin, China won all the medals of "gold, silver and copper" with absolute advantage, and successfully got the prize of 175,000 rupees.

Competition time: Every January.

Competition venue: Mumbai, India.

The official website: http://techfest.org/competitions/Micromouse

● Video

India
Interantional
Micromouse
Challenge

Fig. A–14　Group photo of International Micromouse Challenge, India

It is particularly worth mentioning that the Luban Workshop team of Chennai Institute of technology in India (see Fig. A–15) that adopted the IEEE International standard equipment TQD-Micromouse-JD presented by China in

2017, won the champion of India domestic competition and the fourth place of the world elite group and won the prize of 5,000 rupees, becoming the star team of International Micromouse Challenge, India. Kasik, teacher of Indian Luban workshop, said that such excellent results of the team of Luban Workshop in Chennai Institute of Technology is the result of the joint efforts of teachers and students of Luban Workshop and the support of enterprises, Tianjin Qicheng Science and Technology Co., Ltd. in the past three years.

Fig. A–15　Group photo of winners in the International Micromouse Challenge, India

8) Egypt International Micromouse Competition

Egypt IEEE Institute of Electrical and Electronics Engineers has developed into one of the most influential international academic and technical organizations now. For more than 30 years, the institute has been promoting and guiding the development and innovation of power electronics technology. This technology includes the effective use of electronic components, the application of circuit theory and design technology, and the development of analysis tools for effective conversion, control and power conditions. Our members include outstanding researchers, practitioners and outstanding prize-winners.

As shown in Fig.A–16, IEEE publicizes Micromouse competition on the home page of its official website. IEEE Conference on Power Electronics and Renewable Energy offers generous prizes for the winners of a high-profile international Micromouse competition. Grand Prize: An equivalent of $1,000, Outstanding Performance Award: An equivalent of $700, Best Innovative Design Award: an equivalent of $500. The teams are open to students from Egypt or international engineering or related majors in Egypt, as well as high school students. Within each group, a maximum of two students are allowed.

Fig. A–16    Egypt International Micromouse Competition

The official website: http://www.ieee-cpere.org/International_Competition.html.

Photos of Egypt International Micromouse Competition as shown in Fig. A–17.

Fig. A–17    Photos of Egypt International Micromouse Competition

# Appendix B

## Advanced-Level Classic Competition Analysis

IEEE International Invitational Competition is difficult, it is a challenging and interesting student competition that enjoys a certain reputation and influence at home and abroad. In terms of technology, the Micromouse competition covers Internet of things application technology, electronic information engineering technology, embedded technology, communication technology, software technology, computer network technology, information security technology, mobile communication technology, computer application technology, applied electronic technology, computer control technology, electromechanical integration technology, automation technology and other professional technologies. It involves skills and comprehensive professionalism in sensor detection, artificial intelligence, automatic control and electromechanical motion parts application and fully displaying the development of higher education and vocational education level and improves the training quality of high-quality and high skilled applied talents of electronic information.

The competition can promote the talent cultivation demand of electronic information industry, and adapt to the development of embedded technology, make the intelligent algorithm optimization and other advanced front-end technologies into the competition content in line with the development of science and technology, and guide electronic information majors to carry out single-chip microcomputer application, embedded technology application, Internet of things technology application, etc. The curriculum construction and teaching reform should promote the development of innovative personnel training mode, enhance the employment competitiveness of students majoring in electronic information, promote the education of innovation and entrepreneurship, strengthen the guidance and service of entrepreneurship, and improve the employment level.

With the development of technology, Micromouse conforms to the development of modern technology. After many years of transformation and

optimization, it has become an excellent practical education platform integrating new technologies such as artificial intelligence, embedded, and intelligent sensing.

For more than 40 years, IEEE has held an international Micromouse competition every year. Since its inception, students from all over the world have participated actively, especially college students in the United States and European countries. Therefore, some universities have also set up elective courses on "Micromouse Design Principles and Production Process", as shown in Fig B–1, Fig. B–2.

Fig. B–1　Professor David was in the competition

Fig. B–2　Masakazu Utsunomiya, Japan

Next, we will take typical maze maps of Micromouse competitions at home and abroad as examples to analysis.

## 1. Analysis of the 30th APEC International Micromouse Competition

On March 16, 2015, the 30th APEC international Micromouse competition was held successfully. Tianjin Qicheng Science and Technology Co.,Ltd. led a joint delegation of Tianjin university students to participate in the competition.

It is a historic turning point and the most significant milestone for Tianjin Micromouse technology to integrate with the world.

This competition is a very successful one. The sensor is transiting from digital type to analog type, and the motion structure is also developing from stepping motor to DC motor.

In this event, there is Micromouse integrating the ground suction fan.

The overall difficulty of the maze used in this competition is relatively balanced, and there are many paths to choose, including long straight path to show the high-speed movement performance, and moderate continuous turning to show the precise control, as shown in Fig. B–3.

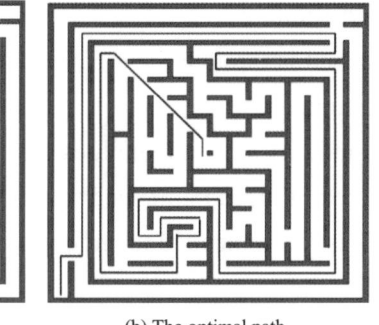

(a) Original maze          (b) The optimal path

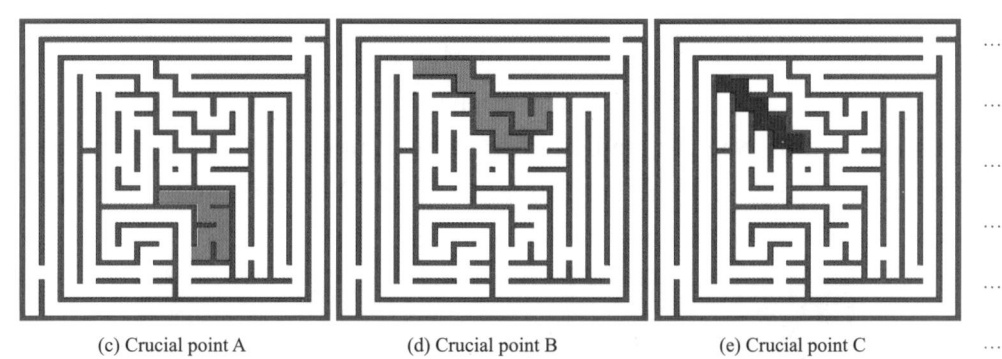

(c) Crucial point A          (d) Crucial point B          (e) Crucial point C

Fig. B–3　Maze analysis

Crucial point A: It is consisting of three T-shaped structures. Combining different directions into more difficult graphics. Not only the accuracy of Micromouse sensor detection, but also the precise control of turning. Due to the large turning frequency and different turning directions, there is no time to correct the posture between turns; once Micromouse has a sensor detection error or a turning angle error, it will be difficult to pass successfully.

Crucial point B: After more than 40 years of development, Micromouse algorithms have been quite intelligent, they are all moving towards the end. Crucial point B is a falsely designed destination path. When Micromouse thinks this is the destination path, once it enters, it must pass through it smoothly. It is also difficult to combine multiple turns and concave paths. It is very important to test the accuracy of infrared detection and turning parameters of Micromouse.

Crucial point C: This is the only way to go to the destination. The difficulty lies in the continuous turning. There are two ways to choose in area C.

The first is to reach the end point through a diagonal line, it can choose to walk along the diagonal line at a 45° turn or turn continuously at 90° to finally reach the end point.

The second is through the area below the straight line and then 90° turning, the final destination.

Both methods have advantages and disadvantages. When walking at 45° can be realized, it is recommended to run the oblique line through the end point; otherwise, it is recommended to choose the second way.

## 2. Analysis of the "Qicheng Cup" Tianjin university student Micromouse competition

This event was hosted by Tianjin Municipal Education Commission, organized by Nankai University and Higher Education Committee of Tianjin Communication Society, and sponsored by Tianjin Qicheng Science and Technology Co.,Ltd. There are 150 teams from 20 colleges and universities participated in the competition, as shown in Fig. B–4.

Fig. B–4　Photo of the competition

The 6th "Qicheng Cup" College Students Micromouse Competition in Tianjin has been successfully held for seven times up to now. The Micromouse developed from a slow stepping motor and ultrasonic sensor to a coreless DC motor and a linear infrared sensor. The level of competitors is also increasing year by year. The 2017 "Qicheng Cup" Micromouse Competition features a classic maze, it is the most important characteristics of high difficulty and openness, as shown in Fig. B-5.

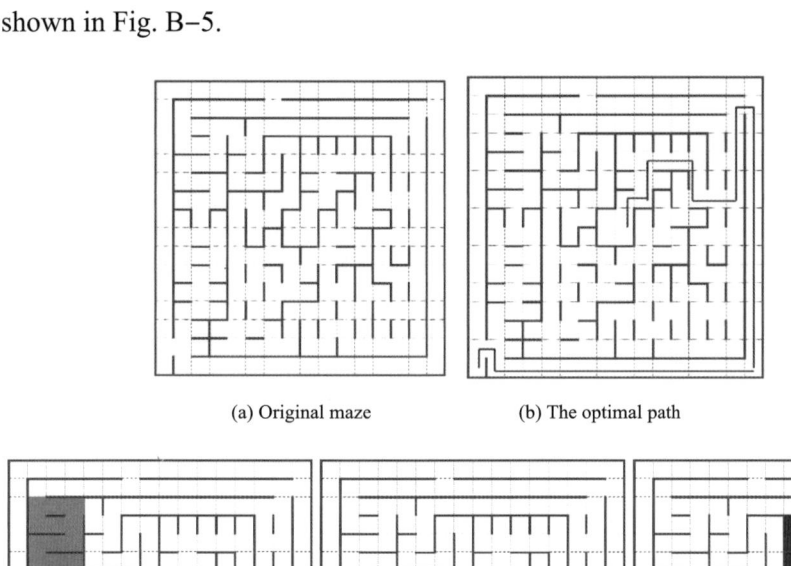

(a) Original maze      (b) The optimal path

(c) Crucial point A      (d) Crucial point B      (e) Crucial point C

Fig. B-5　Maze analysis

Crucial point A: Because the first intersection is close to the start, Micromouse using the intelligent algorithm usually does not turn right here, this resulted in the Micromouse walking all the way around the outermost wall and then coming back to the intersection and finally entering area A. Almost completely symmetrical patterns and numerous intersections make it almost impossible for Micromouse to correct their attitude. This is a great test for the performance of Micromouse.

Crucial point B: Compared with Crucial point A, area B is also highly

open. There are so many intersections that Micromouse do not have enough time for attitude correction. At the same time, it gives high-performance Micromouse the opportunity to run 45° diagonals. It is easy to judge the level of competitors because of the performance gap of Micromouse.

Crucial point C: Area C is the only way to enter the destination. After entering this area, Micromouse using intelligent algorithms usually do not to run "dead end" and directly enter the destination to complete the competition. It also becomes the area to distinguish whether the algorithm is intelligent or not.

### 3. Analysis of the second IEEE Micromouse International Invitational Competition

In May 2017, the second "IEEE Micromouse International Invitational Tournament" were successfully held in China. This competition has attracted representatives from Singapore, Thailand, Mongolia and other countries, as well as Micromouse  elites from Tianjin University, Nankai University, Beijing Jiaotong University, Tianjin Sino German University of Applied Sciences and so on, as shown in Fig. B-6.

Fig. B-6    Photo of the competition

This competition uses a more characteristic maze. With the popularity of intelligent algorithm, the maze selected for large-scale competition pays more and more attention to increase the difficulty coefficient for intelligent algorithm, as shown in Fig. B-7.

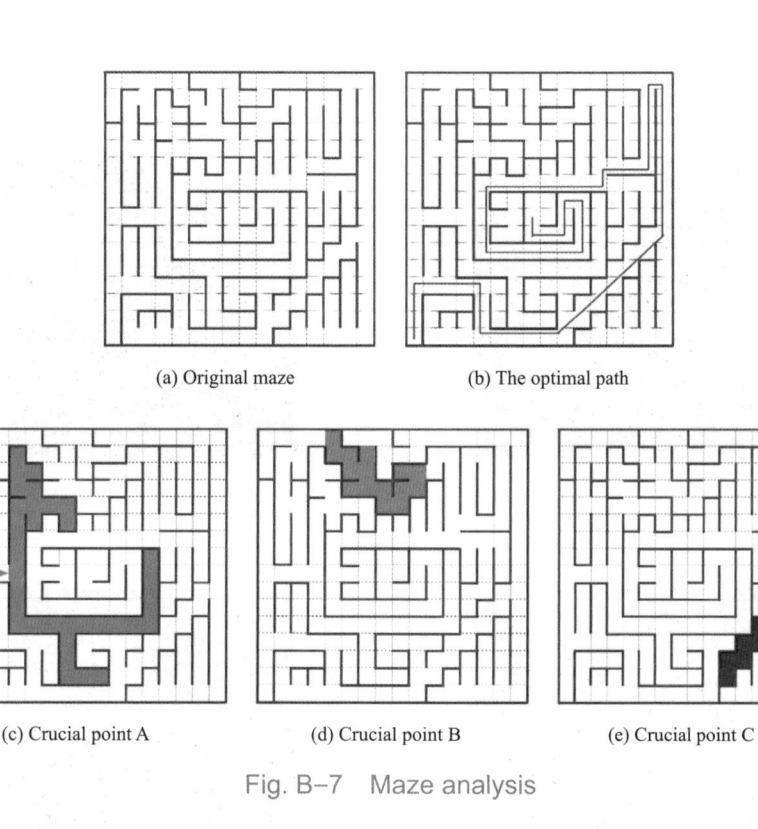

(a) Original maze       (b) The optimal path

(c) Crucial point A       (d) Crucial point B       (e) Crucial point C

Fig. B-7   Maze analysis

Crucial point A: The location of the blue arrow is the preferred intersection for intelligent algorithm. This part is a completely closed area with low difficulty, but when Micromouse enters this area, it needs to face a lot of useless searches and turns. Only after the successful search and exit from area A, Micromouse can enter other regions to search.

Crucial point B: Area B contains two paths for Micromouse to reach the destination. Including a large number of intersections and the combination of left turn and right turn. It pays great attention to the accuracy of infrared detection and turning angle. It will be fatal for Micromouse to misjudge sensors or make wrong turning accuracy at any intersection.

Crucial point C: Area C is also a path to the destination. A series of step-like turns are a great test for the turning accuracy of Micromouse. In the case of 45° oblique straight line running, this path is recommended.

## 4. Maze paradigms of International Micromouse Competition (see Fig. B–8-Fig. B–10)

Fig. B–8   All Japan Micromouse Contest, 2012（expert competition and junior competition）

Fig. B–9   UK International Micromouse Invitational Competition, 2000

Fig. B–10   American APEC International Micromouse Competition, 2002

# Appendix C

## Device List of TQD-Micromouse-JM Ⅱ

Device list of TQD-Micromouse-JM Ⅱ is shown in Table C–1.

Table C–1　Device List of TQD-Micromouse-JM Ⅱ

| No. | Name | Quantity | Remarks |
| --- | --- | --- | --- |
| 1 | TQD-Micromouse-JM Ⅱ | 1 | |
| 2 | Charger | 1 | |
| 3 | Downloader | 1 | |
| 4 | Connecting line | 1 | |
| 5 | USB line | 1 | |
| 6 | Battery | 1 | |
| 7 | Disk | 1 | |

# Appendix D

## Teaching Content and Class Arrangement

The reference teaching hours are 60, and the allocation is shown in Table D–1.

Table D–1　Teaching content and class arrangement

| No | | Teaching Contents | Class hours allocation |
| --- | --- | --- | --- |
| Chapter 1　Elementary Knowledge | Project 1 | Evolution of Micromouse | 22 |
| | Project 2 | Micromouse Hardware Structure | |
| | Project 3 | Development Environment of Micromouse | |
| | Project 4 | Basic Function Control of Micromouse | |
| Chapter 2　Comprehensive Practice | Project 1 | Motion Attitude Control of Micromouse | 18 |
| | Project 2 | Intelligent Control Algorithm and Technology | |
| Chapter 3　Advanced Skills and Competitions | Project 1 | Acquisition and Storage of Maze Information | 20 |
| | Project 2 | The Principle of Path Planning | |
| | Project 3 | Micromouse Program Design | |
| Total | | | 60 |

# Appendix E

## The Circuit Diagram Symbol Comparison Table

The circuit diagram symbol comparison table is shown in Table E–1.

Table E–1   The circuit diagram symbol comparison

| NO. | Name | Drawing methods under China national standard | Drawing methods in software |
|---|---|---|---|
| 1 | Light-emitting diode | | |
| 2 | Ground connection | | |
| 3 | Button switch | | |
| 4 | Diode | | |
| 5 | IGFET | | |

# Appendix F

## Bilingual Comparison Table of Glossary

Bilingual comparison table of glossary as shown in Table F–1 to Table F–3.

Table F–1　Related to Micromouse

| English | Chinese | English | Chinese |
|---|---|---|---|
| main control module | 核心控制模块 | PWM signal generator driver module | PWM信号发生器模块 |
| main control chip | 主控芯片 | the left, the left-oblique, the front, the right-oblique and the right | 左方、左斜、前方、右斜、右方 |
| input module | 输入模块 | The g segment | g段 |
| output module | 输出模块 | coreless DC motor | 空心杯直流电动机 |
| main control circuit | 核心板电路 | stepping motor | 步进电动机 |
| power circuit | 电源电路 | motor drive circuit | 电动机驱动电路 |
| control circuit | 控制电路 | truth table | 真值表 |
| peripheral circuit | 外围电路 | H-bridge circuit | H桥电路 |
| keyboard-display circuit | 键盘显示电路 | rotate | 转动（步进电动机） |
| JTAG interface circuit | JTAG接口电路 | electronic components | 电子元器件 |
| key-pressing circuit | 按键电路 | crystal oscillator | 晶振 |
| data transmission | 数据传输 | capacitance | 电容 |
| human-computer interaction system | 人机交互系统 | adjustable current-limiting resistance | 限流可调电阻 |
| suction fan technology | 吸地风扇技术 | Digitron | 数码管 |
| duty cycle | 占空比 | peripheral devices | 外围器件 |
| angular velocity | 角速度 | pulse oscillation circuit | 脉冲振荡电路 |
| IRsensor | 红外线传感器 | pulse signal | 脉冲信号 |
| infrared detection circuit | 红外检测电路 | square wave | 方波 |
| infrared light | 红外线 | perceptual system | 感知系统 |
| infrared calibration | 红外校准 | carrier frequency | 载波频率 |
| infrared intensity | 红外强度 | schematic Diagram | 原理图 |
| infrared transmitter | 红外发射头 | software interface | 软件界面 |
| infrared receiver | 红外接收头 | driver library | 驱动库 |

Table F-2　Related to competition

| English | Chinese |
| --- | --- |
| cell | 单元格 |
| wall | 挡板 |
| post | 立柱 |
| competition maze | 竞赛场地 |
| the start | 起点 |
| the destination | 目的地/终点 |
| the coordinate in the maze | 迷宫坐标 |
| crossing | 路口 |
| electronic automatic scoring system | 电子自动计分系统 |
| competitor | 参赛队员 |
| micromouse competition | 智能鼠竞速比赛 |
| the optimal path | 最优路径 |
| trajectory | 轨迹 |
| passage way | 通道 |

Table F-3　Related to intelligence algorithm

| English | Chinese | English | Chinese |
| --- | --- | --- | --- |
| the bottom driver program | 底层驱动 | differential-speed control | 差速控制 |
| the top algorithm program | 顶层算法 | straight movement | 直线运动 |
| algorithm | 算法 | turning | 转弯 |
| strategy | 策略 | correct the attitude | 校正车姿 |
| rule | 法则（左、右手法则） | attitude correction | 运行校正 |
| the right-hand rule, the left-hand rule, the central rule | 右手、左手、中心法则 | core function | 核心函数 |
| 90-degree turning/180-degree turning | 90°、180°转弯 | time sequence status | （驱动步进电动机的）时序状态 |
| programming and realizing | 编程并实现 | moving forward one cell | 前进一格 |
| step map | 等高图 | waiting for button press | 按键等待 |
| cycle detection | 循环检测 | determining the attitude | 判断车姿 |
| movement control in picture-8-shaped path | "8字形"路径运行控制 | waiting one step | 暂停一步 |
| obstacle avoidance | 实现避障 | accurate turning control | 精确转弯控制 |
| motion attitude control | 运动姿态的控制 | closed-loop control | 闭环控制 |
| two-wheel difference speed | 两轮差速 | absolute direction | 绝对方向 |
| path planning and decision algorithm | 路径规划和决策算法 | relative direction | 相对方向 |
| struct | 结构体 | | |

# Appendix G
## The International Curriculum Standard for "Micromouse Design Principles and Production Process"

(Applicable to training courses for undergraduates)

## 1. Curriculum Orientation and Characteristics

Applicable Majors: The basic specialties and special specialties in electrical (0806), electronic information (0807), automation (0808) and computer (0809).

Applicable objects: Serve for undergraduate, application-oriented undergraduate and international students of related majors.

Key knowledge points: analog/digital circuit, sensor, embedded hardware technology, embedded programming technology, automatic control technology, advanced intelligent algorithm design, etc.

The orientation and characteristics of this course are as follows:

1) In Line with the *National Standard of Undergraduate Professional Teaching Quality of General Institutions of Higher Learning* issued by the Ministry of Education

Based on the study of general courses and professional basic courses, this course provides a comprehensive teaching carrier for the theory and practice links such as professional model courses, engineering design and curriculum design of senior students, highlights the characteristics of cross integration of "new engineering" disciplines and specialties, highlights the student-centered, focuses on stimulating students' learning interest and potential, and innovates forms and reforms. Reform teaching methods and strengthen practice, and promote the transformation of undergraduate teaching from "good teaching" to "good learning".

2) Connecting with the international teaching concept and serving the cultivation of skilled talents

On the basis of conforming to Chinese standards, the course content is connected with IEEE international competition content, and the design concept of

"application, project and task" is embodied in the curriculum design of teaching materials. The internationally popular ideas of task orientation, systematization of working process and innovation of engineering practice are adopted to enable students to learn professional knowledge, and at the same time, communicate and cooperate with each other stage goals and other forms to improve the comprehensive quality of students.

3) Serve "One Belt and One Road" initiative, spread China's educational standards

The content of this course is highly integrated with the "Luban Workshop" construction projects in many countries. As China's new name card, this course serves "the Belt and Road" initiative, which provides China's educational standards and provides rich practical teaching resources for all countries along the belt, serving the training of skilled personnel in various fields.

## 2. Course objectives

This course takes the professional basic course as the forerunner. Through the learning and training of three progressive models and ten project units, the students will have the engineering literacy of intelligent robot hardware design and drive, software design and programming, project engineering implementation, etc. At the same time, they will master the embedded system hardware design, the internal resource programming and configuration of mainstream microprocessors, the construction of integrated development environment, and the embedded system knowledge and skills of artificial intelligence programming and debugging of equipment. It could help students' in-depth understanding of sensor and detection signal debugging, motor precision control, robot intelligent search and path planning and other professional knowledge.

1) Knowledge objectives

Master the composition of embedded hardware system, the characteristics of embedded microprocessor and memory structure. Master the signal processing and analysis methods of sensors. Master the precise control methods of coreless DC motor. Master the relevant knowledge of realizing intelligent search and path planning in maze.

2) Capability objectives

Master the use methods of common instruments and meters. Master the basic

principles and methods of Micromouse design, assembly and debugging. Master the construction and basic use methods of embedded development environment. Master the method of program design for the overall project. Master the common skills and troubleshooting methods of Micromouse equipment debugging. Have the ability to independently collect and sort out data and information. Have the ability to formulate and implement work plans. Have the ability to explore and use theoretical knowledge to solve practical problems.

3) Quality objectives

Have the ability of communication and teamwork. Have the working style of innovation, dedication and joy, and the craftsman spirit of preciseness, refinement and realism. Have the awareness of safety, quality and responsibility.

## 3. Teaching Content and Requirements

Teaching contents and requirements as shown in Table G–1.

Table G–1　Teaching contents and requirements

| No. | Contents | Teaching requirements | Recomm-ended class hours | Key and difficult points |
|---|---|---|---|---|
| 1 | Chapter 1　Project 1 Evolution of Micromo-use | Understanding the origin of Micromouse. Be familiar with the competition and debugging environment of Micromouse | 4 | Debugging environment of Micromouse |
| 2 | Chapter 1　Project 2 Micromouse Hardware Structure | Master the hardware structure of Micromouse and the relationship between various parts; Understand the selection principles and methods of key devices of Micromouse; Deeply understand the working principle of core circuit of Micromouse | 6 | Working principle of Micromouse core circuit |
| 3 | Chapter 1　Project 3 Development Environment of Micromouse | Familiar with IAR EWARM development environment; Master the program download skills of Micromouse | 2 | IAR EWARM development environment |
| 4 | Chapter 1　Project 4 Basic Function Control of Micromouse | Understand the infrared detection of Micromouse and the working principle of motor drive; Master the attitude detection technology of Micromouse | 8 | Attitude detection technology |

Continued

| No. | Contents | Teaching requirements | Recomm-ended class hours | Key and difficult points |
|---|---|---|---|---|
| 5 | Chapter 2　Project 1 Motion Attitude Control of Micromouse | Master the method of using PID algorithm to stabilize the motor speed; Master the programming realization of Micromouse turning, and master the debugging method of precise control of turning angle | 8 | Principle and implement-ation of PID algorithm |
| 6 | Chapter 2　Project 2 Intelligent Control Algorithm and Technology | Master the methods of controlling Micromouse to run in the middle of the runway and avoid colliding with the wall in front; Master the methods of recognizing and memorizing its own coordinates; Master the control and shape design methods of suction fan | 8 | Walking in the middle, map coordinate representation, wall information recording. |
| 7 | Chapter 3　Project 1 Acquisition and Storage of Maze Information | Master the principle and imple-mentation of relative direction and absolute direction conversion; Master the programming implementation method of common map search strategy | 8 | Principle of maze search optimization rule |
| 8 | Chapter 3　Project 2 The Principle of Path Pla-nning | Master the principle and imple-mentation method of maze search optimization rule; Master the pri-nciple and application method of step map | 8 | Conversion of relative direction and absolute dir-ection |
| 9 | Chapter 3　Project 3 Micromouse Program De-sign | Understand the structure and workflow of the basic program of Micromouse competition | 8 | Adjust the program stru-cture according to different competition strategies |

## 4. Practice teaching

Chapter 1　Project 2　Micromouse PCB Design (2 class hours)

(1) Objective: To learn the hardware design of Micromouse.

(2) Content: Equipment selection and PCB design according to hardware structure diagram.

(3) Steps: ①Equipment selection. ②Determine the wiring requirements of components. ③Design the size and appearance of Micromouse. ④Design and drawing of PCB.

(4) Group: In group of 3-5 person.

(5) Instrument and equipment: Computer, Altium software, key component data manual.

Chapter 1　Project 3　Construction and use of development Environment of Micromouse (2 class hours)

(1) Objective: To build and use learning and development environment.

(2) Content: Install, set up and be familiar with development environment.

(3)Steps: ①Install the IAR development environment. ②Configure the IAR development environment. ③Familiar with the basic operation of the IAR development environment.

(4) Grouping: Independent completion.

(5) Instrument and equipment: Computer, integrated development environment.

Chapter 1　Project 4　Basic Function Control of Micromouse (6 class hours)

(1) Objective: To learn and control Micromouse to complete some basic actions.

(2) Content: Infrared detection, motor drive and attitude detection.

(3) Steps: ①Write infrared driver and sensor information acquisition program. ②Write DC motor driver. ③Write angular velocity sensor information acquisition program.

(4) Group: In group of 3-5 person.

(5) Instrument and equipment: Computer, integrated development environment, Micromouse, LED debugging board.

Chapter 2　Project 1　Motion Attitude Control of Micromouse (6 class hours)

(1) Objective: To learn how to control the motor precisely, and how to use the angular velocity sensor to control Micromouse posture precisely.

(2) Content: Micromouse can run straight and turn accurately

(3) Steps: ①Write a program to use PID algorithm to stabilize the speed of the motor. ②Write a program to accurately control the turning angle of Micromouse. ③Write a program to control the precise turning of Micromouse in the process of moving.

(4) Group: In group of 3-5 person.

(5) Instrument and equipment: Computer, integrated development

environment, Micromouse, $3 \times 3$ maze.

Chapter 2　Project 2　Intelligent Control Algorithm and Technology (6 class hours)

(1) Objective: To learn and control Micromouse to avoid obstacles, update and memorize coordinate information, and master the intelligent control of the suction fan.

(2) Content: Micromouse can avoid obstacles automatically, update coordinates, memorize coordinate information, and use fans to improve stability.

(3) Steps: ①Write a program to use infrared sensor to control Micromouse to run in the middle of the runway. ②Write a program to control Micromouse to recognize the obstacles ahead and brake in time. ③Write a program to control Micromouse to advance several squares accurately and remember its own coordinates. ④ Write a program to control the speed of the suction cup.

(4) Group: In group of 3-5 person.

(5) Instrument and equipment: Computer, integrated development environment, Micromouse, $16 \times 16$ maze.

Chapter 3　Project 1　Acquisition and Storage of Maze Information (6 class hours)

(1) Objective: To learn and control the automatic obstacle avoidance and map information of Micromouse.

(2) Content: Micromouse automatically avoids obstacles, recognizes map information, and remembers map information.

(3) Steps: ①Write a program to realize the mutual conversion of relative direction and absolute direction. ②Write a program to use the direction conversion and infrared sensor information to obtain the correct map information. ③Write a program to realize the common map search strategy.

(4) Group: In group of 3-5 person.

(5) Instrument and equipment: Computer, integrated development environment, Micromouse, $16 \times 16$ maze.

Chapter 3　Project 2　Path Planning and Decision Algorithm (6 class hours)

(1) Objective: To learn Micromouse to carry out intelligent map search and find the best way to the end.

(2) Content: Micromouse common search strategy, common algorithm for path optimization.

(3) Steps: ①Programming to achieve the common search strategy. ②Programming to use flood diffusion algorithm to optimize the search program. ③Programming to use step map algorithm to achieve the sprint program.

(4) Group: In group of 3-5 person.

(5) Instrument and equipment: Computer, integrated development environment, Micromouse, 16×16 maze.

Chapter 3　Project 3　Micromouse Competition (6 class hours)

(1) Objective: To understand the structure and workflow of the basic program of Micromouse competition.

(2) Content: System learning Micromouse search, first sprint, full fig. search, second sprint and other programs.

(3) Step: Write a program to connect Micromouse search, first sprint, full map search and second sprint in a reasonable way, and use the sensor combined with the map features to reasonably configure Micromouse workflow.

(4) Group: In group of 3-5 person.

(5) Instrument and equipment: Computer, integrated development environment, Micromouse, 16×16 maze.

## 5. Suggestions on teaching methods

(1) Teachers should arrange and organize teaching activities according to the typical products in their work tasks.

(2) Teachers should prepare the project assignment according to the learning objectives of the project. The project assignment shall specify the content of the teachers' lecture (or demonstration), the requirements of the learners' preview, the overall arrangement of the project, the training time and content of each module, etc. In case of group learning, the requirements for group arrangement and group discussion (or operation) shall also be clearly specified.

(3) Teachers should take learners as the main body to design teaching structure, create a democratic and harmonious teaching atmosphere, and stimulate learners to participate in teaching activities, improve learners' enthusiasm for learning, and enhance learners' confidence and sense of achievement.

(4) Teachers should guide learners to complete the project completely, and integrate relevant knowledge and skills with professional ethics and emotional attitude.

## 6. Assessment and performance evaluation methods

(1) Assessment method: Examination.

(2) Performance evaluation: Theoretical and practical assessment.

The total score is calculated on the 100 point system, which is divided into two parts: the usual score assessment and the final comprehensive assessment. Generally speaking, attendance, project assignment and test account for 30% of the total score in the usual performance assessment, while the comprehensive assessment at the end of the term accounts for 70% of the total score in the theoretical examination and practical operation assessment, of which the theoretical examination accounts for 30% and the practical operation assessment accounts for 40%.

## 7. Links to other courses

### 1) Prerequisite courses and contact

The prerequisite courses mainly include: Computer principle and composition, electronic technology foundation, signal processing, sensor technology and application, microprocessor principle and application, embedded system and design, artificial intelligence principle and application, etc.

### 2) Follow up courses mainly include: Graduation project

## 8. Teaching materials and reference materials

### 1) Recommended teaching materials

王超，高艺，宋立红. 智能鼠原理与制作：高级篇[M]. 北京：中国铁道出版社有限公司，2019.

### 2) Reference materials

[1] 沈健良，贾玉坤，周芬芬，等. STM32F10X系列之ARM微控制器入门与提高[M]. 北京：北京航空航天大学出版社，2013.

[2] 黄智伟，王兵，朱卫华. STM32F 32位ARM微控制器应用设计与实践[M]. 北京：北京航空航天大学出版社，2012.

## 9. Preparation and review

Prepared by: Deng Bei, Song Lihong, Gao Yuan, Dong rui,Chen Likao, Qiu Jianguo

Reviewed by: Gong Wei, Wang Chao, Gao Yi

June 26, 2020